TUTORIAL C

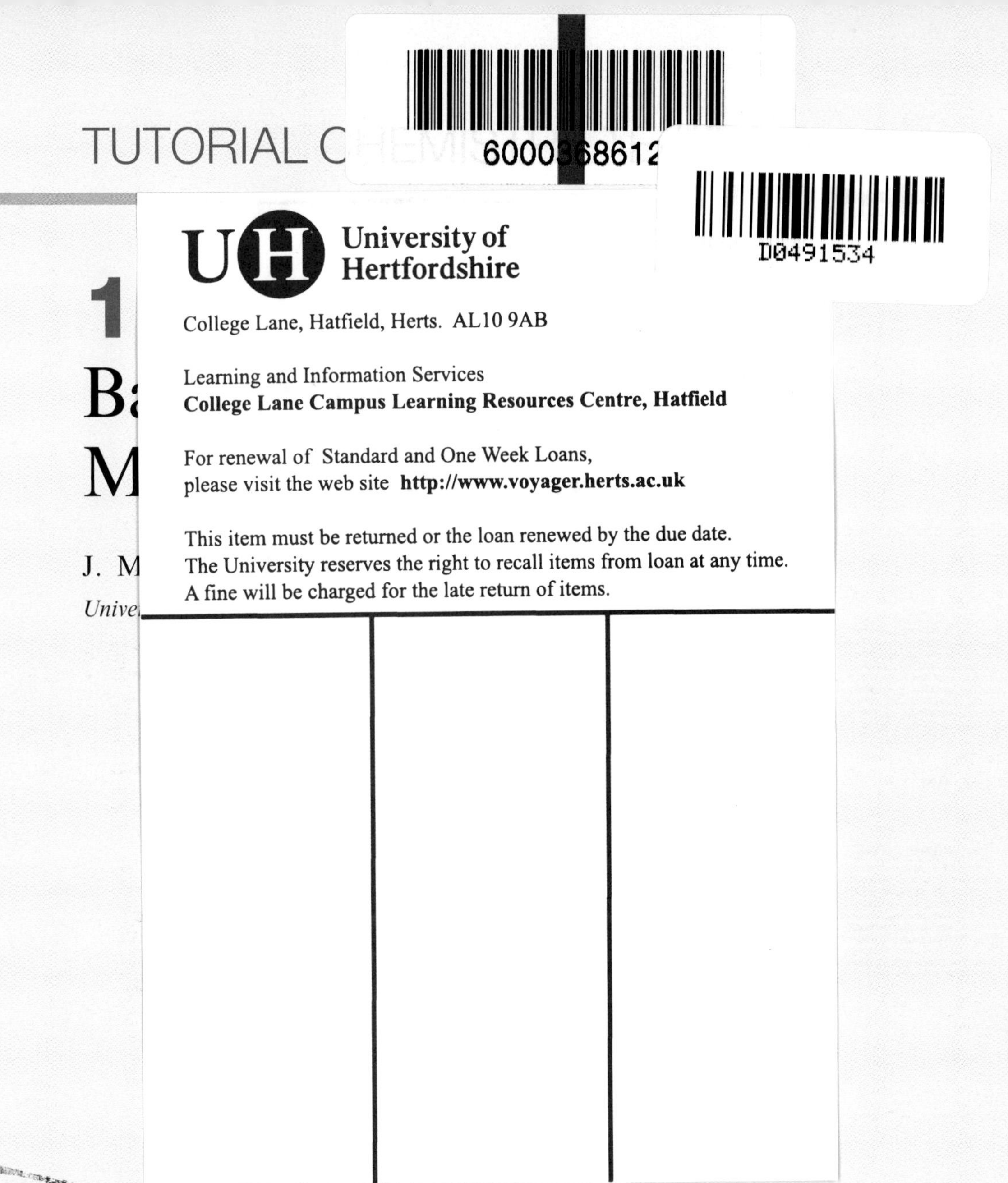

6000368612

D0491534

1

Ba

M



J. M

*Unive*

RS•C

ROYAL SOCIETY OF CHEMISTRY

Cover images © Murray Robertson/visual elements 1998–99, taken from the 109 Visual Elements Periodic Table, available at www.chemsoc.org/viselements



ISBN 0-85404-667-4

A catalogue record for this book is available from the British Library



Published by The Royal Society of Chemistry, Thomas Graham House, Science Park, Milton Road, Cambridge CB4 0WF, UK
Registered Charity No. 207890
For further information see our web site at www.rsc.org

Typeset in Great Britain by Wyvern 21, Bristol
Printed and bound by Polestar Wheatons Ltd, Exeter

# Preface

Spectroscopy is taught at many different levels within the general subject of Chemistry, Physics or Chemical Physics. Before writing the present text I had written two books on spectroscopy: "High Resolution Spectroscopy" is aimed at the postgraduate level, and "Modern Spectroscopy" is aimed at a fairly advanced undergraduate level. Both are referred to as "Further Reading" in association with the present book.

The general level at which the Tutorial Chemistry Texts are aimed is that of the first or second year of a degree course in a British university. To introduce spectroscopy at this level is a considerable challenge. It was my policy during the writing of this book not to be influenced by any existing books (including my own!), but to try to put myself in the position of an undergraduate student early in their course. I assumed that he or she has only limited mathematical knowledge. Consequently, advanced mathematics is avoided as far as possible, appearing only in "boxes" which may be bypassed without sacrificing the understanding of the main text.

It is a truism to say that spectroscopy and quantum mechanics are intimately connected. Quantum mechanics is the theory which attempts to explain, or even to predict, the experimental observations of spectroscopy. However, I have used only the results of quantum mechanics. For many purposes, I want the reader to visualize an atom or molecule as a physical entity rather than a series of mathematical equations.

As a consequence of the various ways in which spectroscopy is commonly encountered, within physical and organic chemistry for example, a student may have the impression that high-resolution (structural) and low-resolution (analytical) spectroscopy can be separated. It has been one of my aims in this book to encompass both, and to show that they are closely interrelated. Another, related, aim has been to integrate the spectroscopy of small and large molecules, thereby trying to avoid the impression that high resolution spectroscopy is mainly concerned with atoms or small, particularly diatomic, molecules.

I would like to express my thanks to Martyn Berry who helped to make the text clearer and more grammatical than it would otherwise have been. I would also like to thank Professor Walter Balfour, for Figures 11.5 and 11.6, and Professor Ben van der Veken, for Figures 10.2 and 10.10, who very kindly provided the original spectra.

J Michael Hollas
*Reading*

# TUTORIAL CHEMISTRY TEXTS



This series of books consists of short, single-topic or modular texts, concentrating on the fundamental areas of chemistry taught in undergraduate science courses. Each book provides a concise account of the basic principles underlying a given subject, embodying an independent-learning philosophy and including worked examples. The one topic, one book approach ensures that the series is adaptable to chemistry courses across a variety of institutions.

TITLES IN THE SERIES

Stereochemistry *D G Morris*
Reactions and Characterization of Solids *S E Dann*
Main Group Chemistry *W Henderson*
d- and f-Block Chemistry *C J Jones*
Structure and Bonding *J Barrett*
Functional Group Chemistry *J R Hanson*
Organotransition Metal Chemistry *A F Hill*
Heterocyclic Chemistry *M Sainsbury*
Atomic Structure and Periodicity *J Barrett*
Thermodynamics and Statistical Mechanics *J M Seddon and J D Gale*
Basic Atomic and Molecular Spectroscopy *J M Hollas*
Organic Synthetic Methods *J R Hanson*
Aromatic Chemistry *J D Hepworth, D R Waring and M J Waring*

FORTHCOMING TITLES

Mechanisms in Organic Reactions
Quantum Mechanics for Chemists
Molecular Interactions
Reaction Kinetics
X-ray Crystallography
Lanthanide and Actinide Elements
Maths for Chemists
Bioinorganic Chemistry
Chemistry of Solid Surfaces
Biology for Chemists
Multi-element NMR

*Further information about this series is available at www.chemsoc.org/tct*

*Orders and enquiries should be sent to:*
Sales and Customer Care, Royal Society of Chemistry, Thomas Graham House, Science Park, Milton Road, Cambridge CB4 0WF, UK

Tel: +44 1223 432360; Fax: +44 1223 426017; Email: sales@rsc.org

# Contents

# Fundamental Constants

| *Quantity* | *Symbol* | *Value and units*[a] |
|---|---|---|
| Speed of light (*in vacuo*) | $c$ | $2.997\,924\,58 \times 10^{8}$ m s$^{-1}$ (exactly) |
| Vacuum permittivity | $\varepsilon_0$ | $8.854\,187\,816 \times 10^{-12}$ F m$^{-1}$ (exactly) |
| Charge on proton | $e$ | $1.602\,176\,462(63) \times 10^{-19}$ C |
| Planck constant | $h$ | $6.626\,068\,76(52) \times 10^{-34}$ J s |
| Molar gas constant | $R$ | $8.314\,472(15)$ J mol$^{-1}$ K$^{-1}$ |
| Avogadro constant | $N_A$ or $L$ | $6.022\,141\,99(47) \times 10^{23}$ mol$^{-1}$ |
| Boltzmann constant | $k\ (= RN_A^{-1})$ | $1.380\,650\,3(24) \times 10^{-23}$ J K$^{-1}$ |
| Atomic mass unit | $u\ (= 10^{-3}$ kg mol$^{-1}$ $N_A^{-1})$ | $1.660\,538\,73(13) \times 10^{-27}$ kg |
| Rest mass of electron | $m_e$ | $9.109\,381\,88(72) \times 10^{-31}$ kg |
| Rest mass of proton | $m_p$ | $1.672\,621\,58(13) \times 10^{-27}$ kg |
| Rydberg constant | $R_\infty$ | $1.097\,373\,156\,854\,8(83) \times 10^{7}$ m$^{-1}$ |

[a] Values taken from P Mohr and B. N. Taylor, *J. Phys. Chem. Ref. Data,* 1999, **28**, 1715 and *Rev. Mod. Phys.,* 2000, **72**, 351. The uncertainties in the final digits are given in parentheses.

# 1
# What is Spectroscopy?

**Aims**

In this short, introductory chapter the reader is introduced to various definitions in spectroscopy, to dispersion and resolution and to the use of units. You should then be able to:

- Understand the words "spectrum", "spectroscope", "spectrograph", "spectrometer" and "spectroscopy"
- Distinguish between dispersion and resolution by a prism or diffraction grating
- Use units of physical quantities with care

## 1.1 What is a Spectrum?

The word **spectrum** is often used in everyday conversation. For example, we may speak of a spectrum of opinions, meaning a range or spread of opinions. This spectrum of opinions may be broad or narrow, depending on the group of people concerned.

In a scientific context, the word spectrum also applies to a range or spread but, in physical chemistry and physics, it applies specifically to a range or spread of **wavelengths** or **frequencies**. An everyday example of a spectrum of frequencies is apparent on our radios. If you look at the various frequency ranges that are available you will see that the range, or spectrum, of FM (frequency modulated) frequencies covered by the radio is typically 88–108 MHz (megahertz). This is a portion of the radiofrequency spectrum. The MW (medium wave) part of the radiofrequency spectrum covers the range 530–1600 kHz (kilohertz). Since the prefixes "mega" and "kilo" indicate $10^6$ and $10^3$, respectively, the FM range is in a higher frequency region of the radiofrequency spectrum than the MW range.

**Worked Problem 1.1**

**Q** Two popular radio programmes are found at 93.7 in the FM range and 1240 in the MW range. At what frequencies, in hertz, do these programmes occur?

**A** In the FM range, 93.7 corresponds to 93.7 MHz which is 93 700 000 Hz. In the MW range, 1240 corresponds to 1240 kHz which is 1 240 000 Hz.

For the purposes of this book, an important example of a spectrum is that of visible light produced by a **prism**. In 1667, Sir Isaac Newton started his experiments on **dispersion** of white light, particularly light from the sun, using a glass prism. He showed that a prism can separate, or **disperse**, white light into its component colours. These are commonly classified as red, orange, yellow, green, blue and violet, although there are no sharp divisions between them.

Figure 1.1 shows how a glass prism can separate white light into its various colours by the process of **refraction**. Light of different colours travels at different speeds through the glass material. Violet light travels most slowly and has the largest **angle of deviation** relative to the beam incident on the prism face, whereas red light travels the fastest and has the smallest angle of deviation. The prism produces a spectrum, the visible part of which consists of the colours to which the eye is sensitive. A rainbow comprises a similar spectrum of colours, and is produced by refraction and reflection of sunlight within the raindrops.

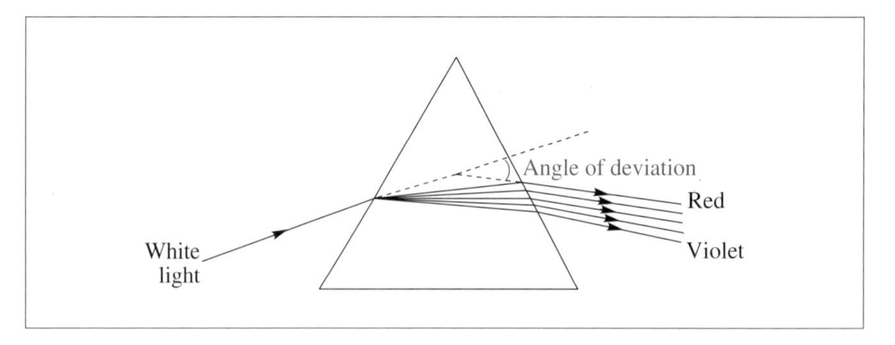

**Figure 1.1** Dispersion of white light by a prism

## 1.2 What is a Spectroscope, Spectrograph, Spectrometer or Spectrophotometer?

As we shall see in Chapter 2, the **electromagnetic spectrum**, as it is called, extends far beyond both the red and the violet, and the various regions of the spectrum are characterized by their wavelengths.

The recommended unit of wavelength is the nanometre (nm), where 1 nm = $10^{-9}$ m. There is another unit of wavelength, the ångström (Å), where 1 Å = $10^{-10}$ m, which has been superseded by the nanometre. However, familiarity with it is useful when reading some of the older literature. The extent of the visible region of the electromagnetic spectrum is approximately 350 nm (violet) to 750 nm (red), or 3500 to 7500 Å.

The simple prism in Figure 1.1 separates the wavelengths only very crudely, and it was not until 1859 that Bunsen and Kirchhoff invented the **spectroscope** in order to analyse the light falling on the prism more accurately. They were then able to use this instrument for chemical analysis.

Figure 1.2 shows the optical arrangement in a typical spectroscope. The light source is behind the entrance slit S1. The light from S1 passes through the lens L1 to the prism which, as in Figure 1.1, disperses the light. Figure 1.2 shows just two light rays 1 and 2, of different colours, with the exit slit S2 positioned so that only ray 1 passes through to be detected by the eye. To direct other colours through S2, the prism is rotated about an axis perpendicular to the figure and through the centre of the prism.

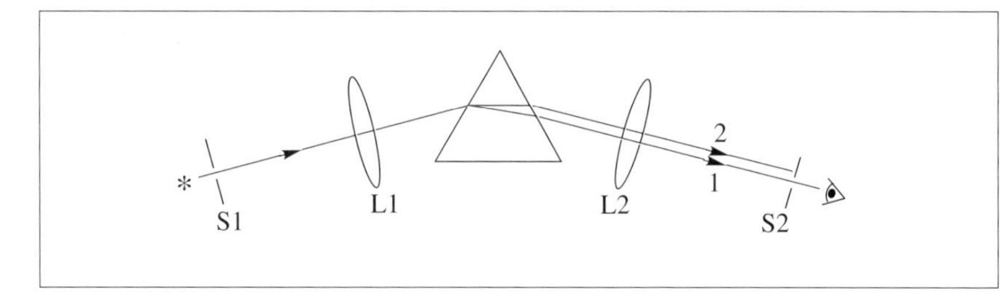


**Figure 1.2** The optical arrangement in a simple prism spectroscope

Even this simple visible-light spectroscope can demonstrate two principles which are important throughout spectroscopy:

1. If the wavelengths of the two rays in Figure 1.2 are $\lambda$ and $\lambda + d\lambda$, where $d\lambda$ is a very small increment of wavelength, and the two rays can be just observed separately, the two wavelengths are said to be resolved and $d\lambda$ is the **resolution** of the spectroscope. A small value of $d\lambda$ corresponds to **high resolution** and a large value to **low resolution**. In fact the resolution of which a glass prism is capable varies with wavelength: it is higher in the violet than in the red region.
2. The eye sees the radiation emerging from the slit S2 as a **spectral line**, and this line is an image of the entrance slit S1. Therefore if the slit S1 is too wide it will restrict the resolution. Ultimately, though, the resolution is determined by the size of the prism. In particular, the resolution is proportional to the length of the base of the prism,

provided the face of the prism on which the light is incident is filled with the source radiation. If the source of radiation is very weak, it may be necessary to widen the slit S1 in order to let through sufficient radiation to be detected. Then the resolution may be **slit limited** and not limited by the size of the prism.

### Worked Problem 1.2

**Q** The **resolving power** $R$ of any dispersing element, such as a prism, is given by:

$$R = \lambda/\mathrm{d}\lambda \tag{1.1}$$

where $\lambda$ is the wavelength and $\mathrm{d}\lambda$ is the resolution. Calculate the resolving power at a wavelength of 405.643 nm and a resolution of 0.94 Å. What is the resolution at a wavelength of 750 nm?

**A** Remember that 1 Å = $10^{-10}$ m and, since 1 nm = $10^{-9}$ m:

$$R = 405.643 \times 10^{-9}\ \mathrm{m} / 0.94 \times 10^{-10}\ \mathrm{m} = 4300$$

Only two figures are significant in the answer, limited by the fact that the resolution is given to only two figures.

At 750 nm, the resolution is given by

$$\mathrm{d}\lambda = 750\ \mathrm{nm}/4300 = 0.17\ \mathrm{nm}\ (= 1.7\ \text{Å})$$

If the exit slit S2 is replaced by a photographic plate detector, the instrument is called a **spectrograph**, but photographic detection is rarely used. Other types of detector, usually some kind of electronic device which records the intensity of radiation falling on it as a function of wavelength, are more useful, and the instrument is then a **spectrometer** or **spectrophotometer**.

Another major difference between a modern spectrometer and one based on Figure 1.2 is the use of a **diffraction grating** in place of the prism as the dispersing element. A diffraction grating consists of a series of very close, parallel grooves ruled onto a hard glassy or metallic surface. The surface is usually reflecting, so that the grating not only disperses the light falling on it but reflects it as well. Since the resolving power of a diffraction grating is proportional to the number of grooves, to obtain high resolution it is necessary for the grooves to be ruled very close together and to cover a large area.

## 1.3 What are Absorption and Emission Spectra?

From what has been discussed so far it will be apparent that **spectroscopy** is the study of experimentally obtained spectra. In general, these spectra may be of two general types, **absorption** and **emission** spectra.

An aqueous solution of copper(II) sulfate, viewed with a white light source passing through it, appears blue because it **absorbs** some of the visible light. If the light passing through the solution is dispersed by a spectroscope or spectrometer it can be shown that it is the red light which has been absorbed. This is because

$$\text{white light} - \text{red light} = \text{blue light}$$

and red and blue are said to be **complementary colours**.

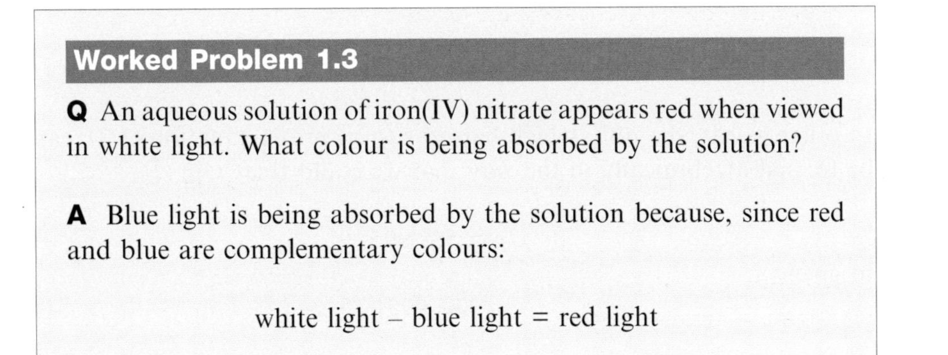

**Worked Problem 1.3**

**Q** An aqueous solution of iron(IV) nitrate appears red when viewed in white light. What colour is being absorbed by the solution?

**A** Blue light is being absorbed by the solution because, since red and blue are complementary colours:

$$\text{white light} - \text{blue light} = \text{red light}$$

An absorption spectrum of great historical importance is that of the sun. The core of the sun acts as a white light source and the absorption is by gases in the outer regions. Wollaston, in 1802, and, later, Fraunhofer observed this spectrum, which consists of sharp, dark **absorption lines**. These lines are images of the narrow slit of the spectroscope or spectrometer, and occur at specific wavelengths which are characteristic of the atom or molecule which is absorbing the light. The strongest absorption lines were identified, by comparison with absorption spectra obtained in the laboratory, as being due to $O_2$, H, Na, Fe, Mg, Ca and $Ca^+$. Of particular interest was an absorption at a wavelength, now measured accurately as 587.5618 nm, which had not been observed in laboratory experiments but which we now know to be due to helium, an element which was first discovered in the sun.

Many spectra are observed in emission rather than absorption. For example, if the radiation emitted from a sodium discharge lamp, similar to those used for street lighting, is directed into a spectroscope, various lines in the emission spectrum can be observed. However, the spectrum is dominated by two intense yellow lines of similar wavelengths, 589.592

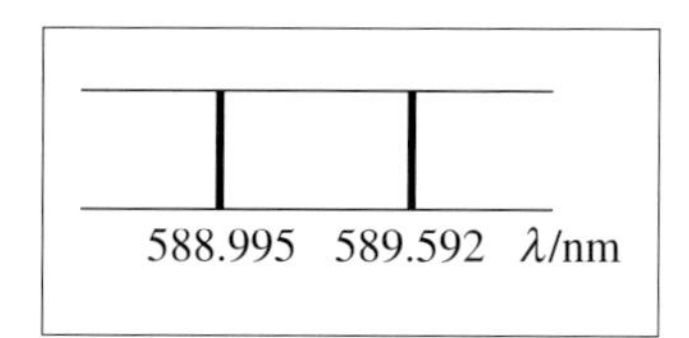


**Figure 1.3** The D lines in the emission spectrum of sodium

and 588.995 nm, known as the sodium D lines. Figure 1.3 shows how these lines might appear to the eye in a spectroscope, or on a photographic recording made with a spectrograph. The width of each line is a consequence of the width of the entrance slit.

There is a rather important principle illustrated by using $\lambda$/nm for labelling the wavelength scale in Figure 1.3. This symbolism indicates that the wavelengths are given in units of nanometres. You might have expected the labelling to be $\lambda$ (nm) and, indeed, you will often find this in books and journals. However, the style used in Figure 1.3 is that which is generally recommended. The logic of this is apparent when we see that, from Figure 1.3 for example:

$$\lambda/\mathrm{nm} = 588.995 \tag{1.2}$$

Multiplying both sides by the unit, nm, gives:

$$\lambda = 588.995\ \mathrm{nm} \tag{1.3}$$

In other words, the unit, which happens to be nm in equation (1.2), can be treated algebraically in the way that we could treat $y$ in:

$$x/y = 588.995$$

to give:

$$x = 588.995y$$

If a d.c. voltage is applied between two electrodes made from iron by bringing them into contact, and the two electrodes are then moved a few millimetres apart, a brilliant bluish-white **arc** will be struck between them. Dispersion of this visible radiation shows that there are hundreds of lines in the emission spectrum of the Fe atom. Indeed, there are so many, spread fairly evenly throughout the spectrum, that their wavelengths have often been used to calibrate other spectra. If two carbon electrodes are used, a brilliant white arc is produced. Although this is similar in colour to that of the iron arc, dispersion shows a very different emission spectrum. The principle species responsible for the emission are the diatomic molecules $C_2$ and CN, the nitrogen originating from that present in the air. Spectra of diatomic molecules are characterized by **bands** of very close, regularly spaced lines rather than the more widely separated, and more randomly spaced, lines which are typical of an atomic spectrum. One of the purposes of this book will be to explain why atomic and molecular spectra are so different in their appearance.

## Summary of Key Points

**1.** *Spectrum, spectroscopy, etc.*
Introduction to the meanings of the words spectrum, spectroscope, spectrometer, spectrophotometer and spectroscopy.

**2.** *Spectral line, resolution*
Observation of a spectral line in a spectroscope. Definition of resolution. Meaning of low and high resolution.

**3.** *Prism, diffraction grating*
Introduction to the use of a prism or diffraction grating to disperse light.

**4.** *Absorption and emission spectra*
Differences between the absorption and emission of light. Examples of both types.

**5.** *Units*
The correct way of expressing units as, for example, in $\lambda$/nm, which is to be preferred to $\lambda$ (nm), to indicate that the wavelength $\lambda$ is in units of nanometres.

## Problems

**1.1.** Two lines in the emission spectrum of calcium (Ca) appear at wavelengths of 443.567 and 443.495 nm. What is the minimum value of the resolving power of a prism or diffraction grating necessary to resolve these two lines?

**1.2.** How would you label the axes of a graph of (a) length, $l$, in centimetres *versus* time, $t$, in minutes and (b) frequency, $\nu$, in kilohertz *versus* magnetic field, $H$, in Gauss (G)?

# 2 The Electromagnetic Spectrum

**Aims**

In this chapter the reader will be introduced to the whole of the electromagnetic spectrum. By the end of this chapter you should:

- Be aware of what lies beyond the red and violet regions
- Understand the difference between wavelength, frequency and wavenumber
- Be able to interconvert units of wavelength, wavenumber, frequency and energy

## 2.1 What Lies Beyond the Red Region?

In Chapter 1 a general discussion of spectroscopy considered only the visible region, from the longer wavelength red region to the shorter wavelength violet region. In fact, as Figure 2.1 shows, the visible region constitutes only a relatively minor portion of the **electromagnetic spectrum**. The reason why this region assumes such great importance is because the human eye is uniquely sensitive to it. However, spectroscopy in general is concerned with all regions of the spectrum.

Figure 2.1 shows that the electromagnetic spectrum stretches far beyond the red region. The **infrared** (IR) is an extensive region which is commonly subdivided into three parts. In the longer wavelength direction these are, in order, the **near-**, **mid-** and **far-infrared** regions. At even longer wavelengths, beyond the infrared, lies the **millimetre wave** region, which partly overlaps the far infrared, followed by the **microwave** region and then the **radiofrequency** region, which includes not only the region used for radio transmission but also, at the shorter wavelength end, that used for television.

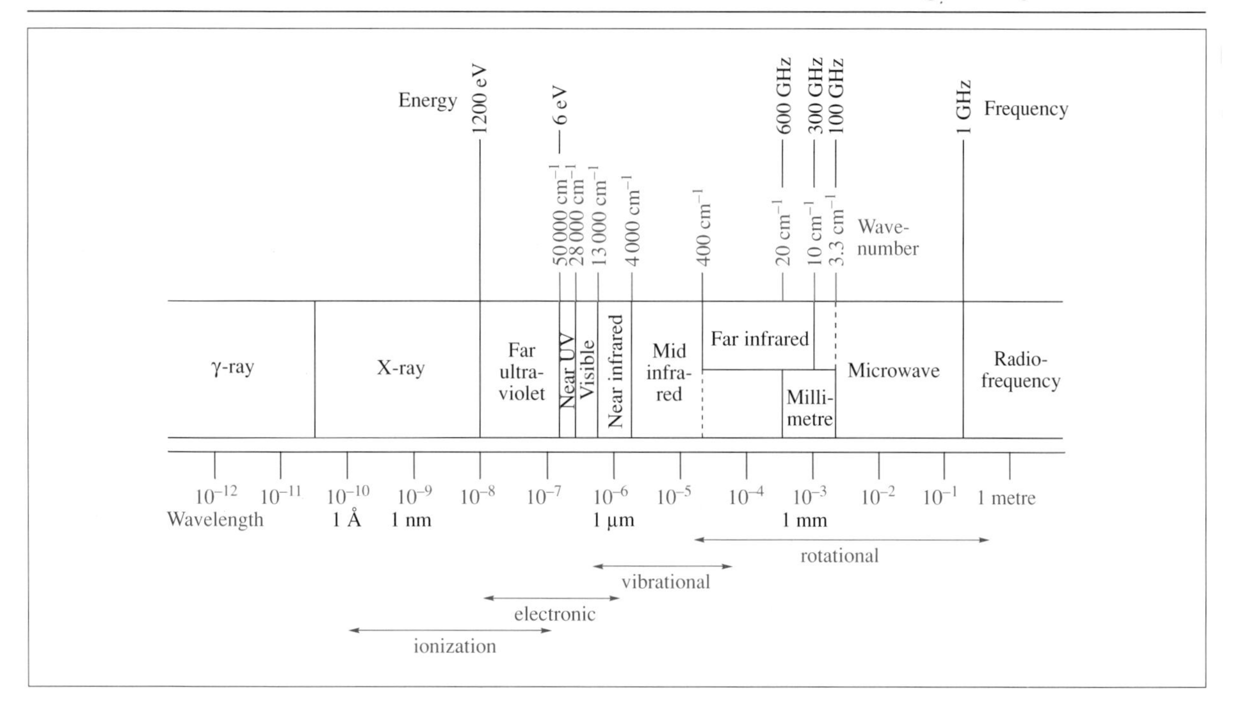


**Figure 2.1** The electromagnetic spectrum

## 2.2 What Lies Beyond the Violet Region?

Figure 2.1 shows that there are four regions that lie to shorter wavelength than the violet region of the electromagnetic spectrum. The **ultraviolet** (UV) region is subdivided into the **near ultraviolet** and the **far ultraviolet**. The far ultraviolet is sometimes called the **vacuum ultraviolet** region, because experiments using it have to be performed in a vacuum.

At even shorter wavelengths lie the **X-ray** and **γ-ray** regions.

## 2.3 Why Electromagnetic?

All **electromagnetic radiation**, in whatever region of the spectrum it falls, can be thought of as consisting of small packets of energy, or **photons**, which can behave both like particles or **waves**. It is the wave-like behaviour which readily illustrates the dual **electric** and **magnetic** character of the radiation.

Figure 2.2 shows the wave-like character associated with a single photon of radiation. The electric and magnetic character are represented by two waves which are **in-phase** and in mutually perpendicular planes. One wave represents an oscillating electric field of strength $E_y$ in the $xy$-plane, and the other an oscillating magnetic field of strength $H_z$ in the $xz$-plane. The wave (or photon) is travelling in the $x$-direction.

When a photon in the visible region falls on the human eye, it is the interaction of the electric component with the retina which results in

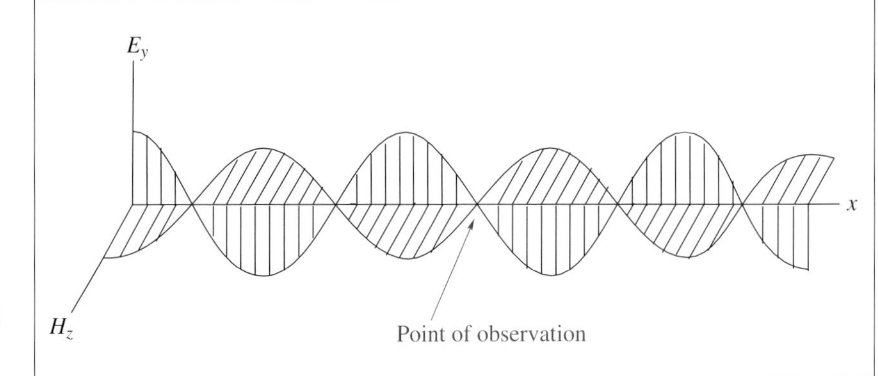


**Figure 2.2** The electric and magnetic components associated with a single photon acting as a wave

detection. It is also the electric character of electromagnetic radiation which is most commonly involved in spectroscopy. NMR (nuclear magnetic resonance) and ESR (electron spin resonance) spectroscopies are important exceptions, but will not be considered in this book.

All electromagnetic radiation travels, in a vacuum, with the **speed of light**, $c$, which has the exactly defined value:

See page viii for a list of fundamental constants.

$$c = 2.997\,924\,58 \times 10^8 \text{ m s}^{-1} \tag{2.1}$$

When radiation is **polarized**, this usually refers to **plane polarization**, and it is conventional to refer to the plane of the oscillating electric field. In the case illustrated in Figure 2.2, the polarization is in the $xy$-plane.

## 2.4 Units of Wavelength, Frequency, Wavenumber and Energy

In Figure 2.1 the **wavelength** of radiation is expressed primarily in terms of the base unit of length, the **metre**. However, for general use, it is much more convenient to use units derived from the metre using prefixes indicating various powers of ten. For units in general the prefixes most frequently encountered are:

$10^{12}$ tera (T), $10^9$ giga (G), $10^6$ mega (M), $10^3$ kilo (k), $10^{-2}$ centi (c), $10^{-3}$ milli (m), $10^{-6}$ micro (μ), $10^{-9}$ nano (n), $10^{-12}$ pico (p), $10^{-15}$ femto (f) (2.2)

Wavelength can be used to indicate a particular position in the electromagnetic spectrum which, in future, we shall refer to simply as "the spectrum". Indeed, wavelength is often what is measured experimentally in a spectrometer since it is related relatively simply to, say, the angle of rotation of a prism or of a diffraction grating used in the spectrometer.

**Worked Problem 2.1**

**Q** (a) Write down the full names for the units of (i) mm, (ii) μm and (iii) nm.
(b) What is the derived unit for $10^{-12}$ m?
(c) Use prefixes to express the units for (i) $1.3 \times 10^{-5}$ s and (ii) $0.27 \times 10^{-14}$ s.

**A** (a) (i) millimetre, (ii) micrometre and (iii) nanometre.
(b) $10^{-12}$ m = 1 pm (picometre).
(c) (i) 13 μs (microsecond), (ii) 2.7 fs (femtosecond).

Exceptions are microwave and millimetre wave spectroscopy, in which it is the frequency which is measured directly.

In general, when it comes to analysing spectra, it is either frequency or wavenumber which is the most useful.

In Figure 2.2, the meaning of wavelength is clear: it is simply the distance, along the $x$-axis, between adjacent positive (or adjacent negative) peaks. If the wave is travelling in the $x$-direction and observed at any point, as shown, the number of waves which pass per second can be counted. This is then the **frequency**, in units of $s^{-1}$; this unit is usually referred to as the hertz, where 1 $s^{-1}$ = 1 Hz.

Alternatively, the number of waves in 1 m of distance along the $x$-axis can be counted and this gives the **wavenumber** of the radiation in units of $m^{-1}$. It is more usual to express a wavenumber in units of $cm^{-1}$, the number of waves per centimetre, and it is only this unit which we shall use subsequently.

The symbols used for wavelength, frequency and wavenumber are $\lambda$, $\nu$ and $\tilde{\nu}$, respectively. They are related in the following way:

$$\nu = c/\lambda \tag{2.3}$$

where $c$ is the speed of light (see equation 2.1), and:

$$\tilde{\nu} = 1/\lambda \tag{2.4}$$

The wavelength of radiation depends on the medium through which it is travelling, and, in spectroscopy, it is necessary to convert all wavelengths, which are usually measured in air, to vacuum wavelengths using the relation:

$$\lambda_{vac} = n_{air}\lambda_{air} \tag{2.5}$$

where $n_{air}$ is the **refractive index** of air, and is related to the speed of light in the two media by:

$$c_{vac} = n_{air}c_{air} \tag{2.6}$$

Frequency is independent of the medium in which it is measured because:

$$\nu = c_{air}/\lambda_{air} = c_{vac}/\lambda_{vac} \tag{2.7}$$

Wavenumber, however, is not independent of the medium but, in spectroscopy, is invariably taken to be defined as:

$$\tilde{\nu} = 1/\lambda_{vac} \tag{2.8}$$

Strictly, it should be referred to as the vacuum wavenumber but is usually referred to, simply, as the wavenumber.

### Worked Problem 2.2

**Q** The wavelength limits of the visible region of the spectrum are about 350–750 nm. Convert these limits to wavenumber, in units of $cm^{-1}$.

**A** Using equation (2.4), the wavenumber limits are:

$$\begin{aligned}\tilde{\nu} &= 1/350 \times 10^{-9}\ \text{m to } 1/750 \times 10^{-9}\ \text{m}\\ &= 1/350 \times 10^{-7}\ \text{cm to } 1/750 \times 10^{-7}\ \text{cm}\\ &= 28\,600\ \text{cm}^{-1} \text{ to } 13\,300\ \text{cm}^{-1}\end{aligned}$$

The approximate wavenumber limits of the visible region, together with those of the near ultraviolet, near infrared, mid infrared, far infrared and millimetre wave regions, are shown in Figure 2.1 with units of $cm^{-1}$.

In millimetre wave and microwave spectroscopy, frequency rather than wavenumber is usually used, and the approximate limits of these regions are given in Figure 2.1 in units of GHz.

**Worked Problem 2.3**

**Q** The limits of the microwave region are approximately 1 GHz and 100 GHz, and those of the millimetre wave region are approximately 100 GHz and 300 GHz. Convert these frequencies to wavelengths.

**A** From equation (2.3):

$$\lambda = c/\nu = \text{(for 1 GHz) } 3.00 \times 10^8 \text{ m s}^{-1}/1 \times 10^9 \text{ s}^{-1}$$
$$= \text{(for 100 GHz) } 3.00 \times 10^8 \text{ m s}^{-1}/100 \times 10^9 \text{ s}^{-1}$$
$$= \text{(for 300 GHz) } 3.00 \times 10^8 \text{ m s}^{-1}/300 \times 10^9 \text{ s}^{-1}$$
$$= 0.3 \text{ m, } 3 \text{ mm or } 1 \text{ mm}$$

Energy, $E$, is associated with all electromagnetic radiation and is simply related to the frequency or wavenumber of the radiation by:

$$E = h\nu = hc\tilde{\nu} \tag{2.9}$$

where $h$ is the Planck constant, which has the value:

$$h = 6.626\ 068\ 76 \times 10^{-34} \text{ J s} \tag{2.10}$$

The energy given by equation (2.9) is that associated with a single photon of radiation of frequency $\nu$ or wavenumber $\tilde{\nu}$, and is extremely small.

**Worked Problem 2.4**

**Q** What is the energy of one photon of radiation of (a) frequency 4.6 GHz, and (b) wavenumber 37 000 $\text{cm}^{-1}$? What is the energy of a mole of these photons?

**A** (a) $E = h\nu = 6.6 \times 10^{-34} \text{ J s} \times 4.6 \times 10^9 \text{ s}^{-1} = 3.0 \times 10^{-24} \text{ J}$

The Avogadro constant, $N_A$, is the number of particles (photons in this case) in one mole where:

$$N_A = 6.022\ 141\ 99 \times 10^{23} \text{ mol}^{-1} \tag{2.11}$$

$\therefore$ for 1 mole: $E = N_A h\nu = 6.0 \times 10^{23} \text{ mol}^{-1} \times 3.0 \times 10^{-24} \text{ J} = 1.8 \text{ J}$
(b) $E = hc\tilde{\nu} = 6.6 \times 10^{-34} \text{ J s} \times 3.0 \times 10^8 \text{ m s}^{-1} \times 37\ 000 \times 10^2 \text{ m}^{-1}$
$= 7.3 \times 10^{-19} \text{ J}$

$\therefore$ for 1 mole: $E = 6.0 \times 10^{23}\ \text{mol}^{-1} \times 7.3 \times 10^{-19}\ \text{J} = 4.4 \times 10^{5}\ \text{J mol}^{-1} = 440\ \text{kJ mol}^{-1}$

The energy calculated in (b) is about the same as the **dissociation energy** for a single bond. For example, that of the $H_2$ molecule is 436 kJ mol$^{-1}$. Therefore there is just sufficient energy in this near-ultraviolet radiation to dissociate hydrogen molecules into hydrogen atoms: $H_2 \rightarrow 2H$.

Note that, in this problem as elsewhere, it is very useful to insert the units of all the physical quantities used in calculations. This serves to check that the final units are correct, and also to check the calculation.

Another unit of energy which is commonly used, particularly in the far ultraviolet and X-ray regions, is the electronvolt (eV) which is defined as the kinetic energy acquired by an electron which has been accelerated by a potential difference of 1 volt (V). The electronvolt is related to the more usual unit of energy by:

$$1\ \text{eV} = 1.602\ 18 \times 10^{-19}\ \text{J} = 96.485\ \text{kJ mol}^{-1} \qquad (2.12)$$

and is related to wavenumber and frequency by:

$$1\ \text{eV} = hc(8065.54\ \text{cm}^{-1}) = h(2.417\ 99 \times 10^{14}\ \text{s}^{-1}) \qquad (2.13)$$

## 2.5 The Effect of Radiation on Atoms and Molecules

Equation (2.9) shows that, in moving from the radiofrequency to the γ-ray region of the spectrum, the photon energy increases. This is apparent when we contrast the benign nature of radio waves, which are everywhere around us, with the extremely dangerous γ-rays.

Similarly, when atoms or molecules are subjected to radiation they suffer more drastic consequences as the frequency of the radiation increases. For example, Figure 2.1 shows that the energy in the far-ultraviolet and X-ray regions is sufficient to ionize an atom or molecule M by the process:

$$M + h\nu \rightarrow M^{+} + e^{-}\ \text{(an electron)} \qquad (2.14)$$

A typical **ionization energy**, the energy required for this process, is about 10 eV, or 960 kJ mol$^{-1}$.

Figure 2.1 also indicates that radiation in the visible and ultraviolet regions is typically sufficient to promote an electron from a lower to a higher energy orbital in an atom or molecule by an **electronic** process.

Only molecules have **vibrational** and **rotational degrees of freedom**. Vibrational motion in a molecule is excited typically by infrared radiation, while rotational motion requires only the lower energy of far infrared, millimetre wave or microwave radiation.

## 2.6 Subdivisions of Spectroscopy

Figure 2.1 illustrates the two general ways by which the subject of spectroscopy is commonly divided. One of these is referred to by indicating the region of the electromagnetic spectrum involved. For example, we may refer to microwave, infrared, visible or ultraviolet spectroscopy, depending on the type of the source of radiation used in obtaining an absorption spectrum, or the radiation emitted in obtaining an emission spectrum. Clearly these subdivisions are based on the experimental technique employed and, therefore, the type of spectrometer used.

Of much greater fundamental importance in spectroscopy than the nature of the radiation involved is the type of process induced in the atom or molecule which is either subjected to or emits the radiation. The processes with which we shall be concerned in this book, and which have been introduced in Section 2.4, are, in order of increasing energy, rotational, vibrational, electronic and ionization. In this way, spectroscopy can be subdivided into rotational, vibrational, electronic and ionization spectroscopy. One type of ionization spectroscopy, involving the ionization process in equation (2.14), is called **photoelectron spectroscopy**. The electron ejected when the atom or molecule encounters the photon of energy $h\nu$ is referred to as a **photoelectron**.

There is a further type of spectroscopy, involving neither an absorption nor an emission process, which will concern us: that is **Raman spectroscopy** in which radiation is **scattered** by the sample. The source radiation used in Raman spectroscopy may be in the near-ultraviolet, visible or near-infrared region.

### Summary of Key Points

1. *Regions of the electromagnetic spectrum*
In this book, the most important are the microwave, far, mid and near infrared, visible, near and far ultraviolet, and, marginally, the X-ray region.

**2.** *Why it is called "electromagnetic" radiation*
Radiation consists of electric and magnetic components. Concept of polarized radiation.

**3.** *Interrelation between wavelength, frequency, wavenumber and energy*
Frequency and wavenumber of features in spectra are the most often used. Importance of distinguishing between them.

**4.** *Correct use of units*
Interconverting wavelength, frequency, wavenumber and energy requires careful use of units. Insert units of all physical quantities in a calculation to make sure the units of the result are correct.

**5.** *Effects of radiation on atoms and molecules*
These depend on the energy of the radiation. For example, low-energy microwave radiation causes molecules to rotate; high-energy far-ultraviolet or X-ray radiation ionizes them.

**6.** *Subdivisions of spectroscopy*
Subdivision according to either the type of radiation involved or the process taking place in the atom or molecule.

## Problems

**2.1.** Use prefixes, attached to the units, to express in convenient form: (a) $13.5 \times 10^{-7}$ s, (b) $253 \times 10^{-5}$ g, (c) $1743 \times 10^{7}$ Hz, (d) $12.6 \times 10^{-10}$ m.

**2.2.** (a) Convert to wavenumber, with units of $cm^{-1}$, (i) a frequency of 9.74832 GHz, (ii) a wavelength of 6437.846 Å (see Section 1.2 for this unit).
(b) Convert 12.488 eV to energy with units of J and J $mol^{-1}$.

# 3 Quantization and the Hydrogen Atom

**Aims**

In this chapter you will be introduced to:

- Quantization: the discrete character of the energy possessed by an atom or molecule
- Quantization of energy in the hydrogen atom
- The Bohr theory of the hydrogen atom
- The photoelectric effect
- Electrons behaving as waves
- Simple quantum mechanical treatment of the hydrogen atom

## 3.1 What is Quantization?

The D lines in the emission spectrum of atomic sodium, illustrated in Figure 1.3, are an example of what appeared to be, in the earliest days of spectroscopy, a remarkable phenomenon. The yellow colour of the sodium emission would have been expected to cover a broad band of wavelengths in the yellow region of the spectrum. Instead, there are only two lines in this region, with precise, **discrete** wavelengths.

If we think in energy terms, we can see that each of the sodium D lines represents the **energy difference** between two discrete **energy levels** of the sodium atom. It is the discrete character of energy levels in both atoms and molecules which is referred to as **quantization**.

Figure 3.1 shows such a pair of energy levels, $E_1$ and $E_2$, where $E_2 > E_1$. The energy difference, $\Delta E$, is given by:

$$\Delta E = E_2 - E_1 \qquad (3.1)$$

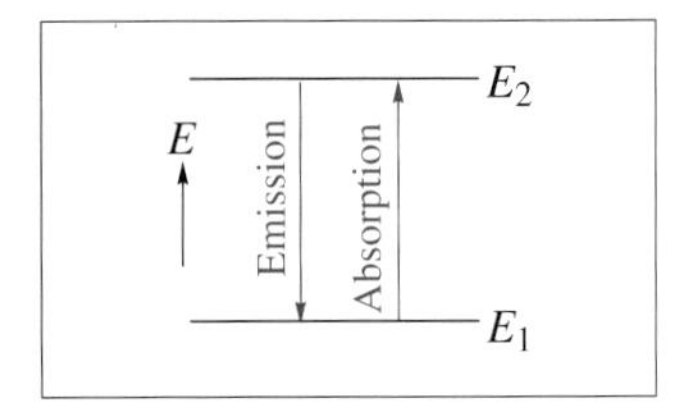


**Figure 3.1** Absorption and emission processes between two discrete energy levels $E_1$ and $E_2$

Figure 3.1 shows that, in general, a **transition** between two energy levels may be an emission or absorption process. Each photon emitted or absorbed is referred to as a **quantum** of radiation.

## Worked Problem 3.1

**Q** Use the wavelengths of each of the sodium D lines given in Figure 1.3 to obtain the corresponding energy differences, $\Delta E$, in units of kJ mol$^{-1}$, for each transition.

**A** For the shorter wavelength line:

$$\lambda = 588.995 \text{ nm} = 588.995 \times 10^{-9} \text{ m}$$

Using equations (2.4) and (2.5) to convert to energy gives:

$$\begin{aligned}\Delta E &= hc/\lambda \\ &= 6.626\,07 \times 10^{-34} \text{ J s} \times 2.997\,92 \times 10^{8} \text{ m s}^{-1}/588.995 \times 10^{-9} \text{ m} \\ &= 3.372\,60 \times 10^{-19} \text{ J}\end{aligned}$$

This is the energy of one quantum, or photon, of radiation of wavelength 588.995 nm.

Alternatively, for one mole of photons:

$$\Delta E = 3.372\,60 \times 10^{-22} \text{ kJ} \times 6.0221\,4 \times 10^{23} \text{ mol}^{-1}$$

(*i.e.* energy of one photon × the Avogadro constant, $N_A$)

$$= 203.103 \text{ kJ mol}^{-1}$$

For $\lambda = 589.592$ nm:

$$\Delta E = 3.369\,19 \times 10^{-19} \text{ J or } 202.897 \text{ kJ mol}^{-1}$$

Note how six figures have been used for all the constants in the calculations, and also in the answers, consistent with the six-figure accuracy to which the wavelengths were given.

## 3.2 Quantization of Energy in the Hydrogen Atom

One of the earliest examples of an experimental observation which showed that an atom can have only discrete amounts of energy was that of the emission spectrum of the hydrogen atom, part of which was observed in the visible region by Balmer in 1885. The spectrum can be obtained by producing an electrical discharge in hydrogen gas at very low pressure. The hydrogen molecules are dissociated into hydrogen atoms by collisions with electrons produced in the discharge. The hydrogen atoms are formed with a wide range of energies, and the Balmer series extends from the visible into the near ultraviolet region; it is illustrated in Figure 3.2.

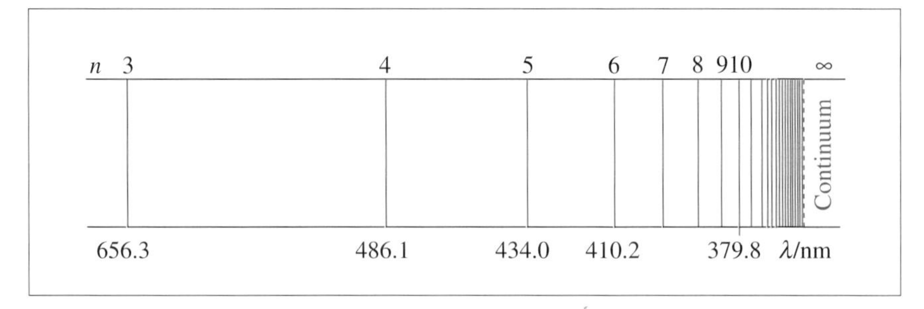


**Figure 3.2** The Balmer series of lines in the emission spectrum of the hydrogen atom

Like the spectrum of sodium in Figure 1.3, this spectrum also shows that the atom emits energy in discrete amounts, giving rise to sharp lines in the spectrum. More importantly, unlike the spectra of nearly all other atoms, it shows a remarkable series of lines converging smoothly to shorter wavelengths. The series is now known as the **Balmer series**. Beyond the convergence limit there are no discrete lines but a **continuum** in which energy is emitted over a continuous range of wavelengths.

Balmer was not able to observe the shorter wavelength region of this spectrum, but the longer wavelength lines that he observed were sufficient to show the smooth convergence. This prompted him to search for a mathematical relationship between the wavelengths, $\lambda$, and a running number, $n'$, labelling each of the lines. The relationship which he obtained is:

$$\lambda = n'^2G/(n'^2 - 4) \tag{3.2}$$

where $G$ is a constant and $n'$ is an integer which can take the values 3, 4, 5, ... for successive lines. For the first line, at 656.3 nm, $n' = 3$.

Equation (3.2) represented an important landmark in spectroscopy. It was the first successful attempt to apply theory to experimental observations, and marked the beginning of a very close relationship between the results of experimental spectroscopy and theory.

Using the relationship between wavelength and frequency in equation (2.3), equation (3.2) can be rewritten as:

$$\nu = R_H(1/2^2 - 1/n'^2) \tag{3.3}$$

where $R_H$ is the Rydberg constant for hydrogen, and is given by:

$$R_H = \mu e^4/8h^3\varepsilon_0^2 \quad (3.4)$$

In this equation the fundamental constants $e$, $h$ and $\varepsilon_0$ are the charge on the electron, the Planck constant and vacuum permittivity, respectively. The quantity $\mu$ is the reduced mass of the system (proton plus electron) given by:

$$\mu = m_e m_p/(m_e + m_p) \quad (3.5)$$

where $m_e$ and $m_p$ are the masses of the electron and the proton, respectively.

In terms of wavenumber, the positions of members of the Balmer series are given by:

$$\tilde{\nu} = \tilde{R}_H\,(1/2^2 - 1/n'^2) \quad (3.6)$$

where:

$$\tilde{R}_H = R_H/c \quad (3.7)$$

**Worked Problem 3.2**

**Q** Calculate the value of $\tilde{R}_H$, with units of $cm^{-1}$ and to six significant figures, using the values of the fundamental constants given on page viii.

**A** From equation (3.5), the reduced mass for the hydrogen atom is given by:

$\mu = 9.109\ 38 \times 10^{-31}$ kg $\times$ $1.672\ 62 \times 10^{-27}$ kg/($9.109\ 38 \times 10^{-31}$ kg + $1.672\ 62 \times 10^{-27}$ kg)
$= 9.104\ 42 \times 10^{-31}$ kg

Note that, because the proton has a very much greater mass than that of the electron, the reduced mass is very similar to, but significantly less than, that of the electron alone.

From equations (3.4) and (3.7):

$\tilde{R}_H = 9.104\ 42 \times 10^{-31}$ kg $\times$ $(1.602\ 18 \times 10^{-19})^4$ $C^4$/8 $\times$ $(6.626\ 07 \times 10^{-34})^3$ $J^3\,s^3$ $\times$ $(8.854\ 19 \times 10^{-12})^2$ $F^2\,m^{-2}$ $\times$ $2.997\ 92 \times 10^{10}$ cm $s^{-1}$
$= 1.096\ 79 \times 10^5$ $cm^{-1}$

In this calculation it is essential to deal with the powers of ten first before multiplying and dividing the other numbers.

The units of the result, $cm^{-1}$, do not follow easily from the units which have been inserted for the various constants. It is sufficient to note here that all the constants, except $c$, have units which are derived from the SI (Système International) base units. Therefore if $c$ were given with units of m $s^{-1}$, the resulting value for $\tilde{R}_H$ would have units of $m^{-1}$. Inserting the value of $c$ with units of cm $s^{-1}$ ensures that $\tilde{R}_H$ has units of $cm^{-1}$.

Equation (3.6) represents the difference between two terms, each of the form $-\tilde{R}_H/n^2$. Conversion of these terms from wavenumber to energy gives:

$$E = -\tilde{R}_H\,hc/n^2 = -(\mu e^4/8h^2c\varepsilon_0^2)/n^2 \qquad (3.8)$$

These are the quantized energy levels of atomic hydrogen, where the **quantum number**, $n$, can take any integral value $n = 1, 2, 3, 4, \ldots \infty$, and are shown in Figure 3.3.

Transitions between the energy levels occur with energies given by:

$$\Delta E = (\mu e^4/8h^2c\varepsilon_0^2)/(1/n''^2 - 1/n'^2) \qquad (3.9)$$

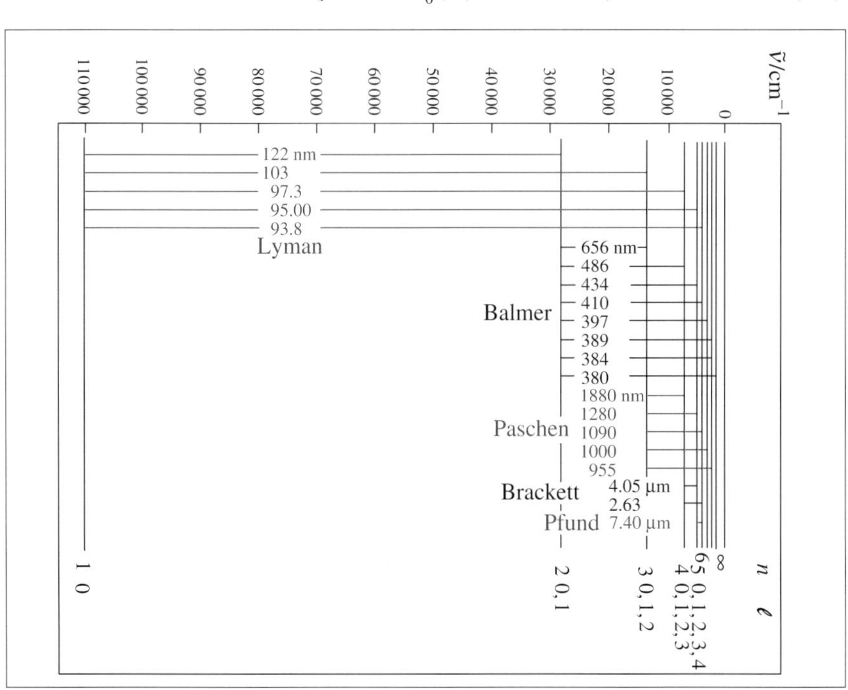


**Figure 3.3** Energy levels and transitions of the hydrogen atom

Double primes (") and single primes (') are used throughout spectroscopy to indicate, respectively, the lower and upper levels, or states, of any type of transition.

or, alternatively, with wavenumbers given by:

$$\tilde{\nu} = \tilde{R}_H (1/n''^2 - 1/n'^2) \tag{3.10}$$

where $n''$ and $n'$ are the quantum numbers for the lower and upper levels, respectively, of a transition.

Figure 3.3 shows that there is an infinite number of series in the spectrum of hydrogen with $n'' = 1, 2, 3, 4, \ldots \infty$. Those series with $n'' = 1$, 2, 3, 4 and 5 have the names of their discoverers, Lyman, Balmer, Paschen, Brackett and Pfund, respectively, attached to them. The series appear in various regions of the spectrum, ranging from the Lyman series, in the far ultraviolet, to series with very high values of $n''$ in the radiofrequency region.

The convergence limit of the energy levels, at $n = \infty$, corresponds to ionization of the hydrogen atom in the process:

$$H + h\nu = H^+ + e^- \tag{3.11}$$

in which a quantum of radiation, $h\nu$, has sufficient energy to remove the electron. Since the ejected electron may have any amount of kinetic energy, the energy levels are no longer discrete but form a continuum beyond the ionization limit.

Members of series with very high values of $n''$ are particularly important in observing the large amount of atomic hydrogen which is found throughout the universe. For example, the transition with $n'' = 166$ and $n' = 167$ has been observed in the regions between stars, the interstellar medium, with a frequency of 1.425 GHz.

Figure 3.3 also illustrates an example of a **selection rule**. It shows that transitions with all values of $\Delta n$ $(= n' - n'')$ are allowed:

$$\Delta n = 1, 2, 3, 4, 5, \ldots \infty \tag{3.12}$$

so that each series consists of an infinite number of lines.

This selection rule is the first of many that we shall encounter in spectroscopy and is unusual in that it is completely unrestrictive.

**Worked Problem 3.3**

**Q** Using six significant figures, calculate the wavenumbers of the transitions in the hydrogen atom with the following values of the quantum number $n$:

| | $n''$ | $n'$ |
|---|---|---|
| (a) | 1 | 5 |
| (b) | 3 | 20 |
| (c) | 102 | 154 |
| (d) | 153 | 154 |

Convert the result for (a) to wavelength and that for (d) to frequency and state the region of the spectrum in which each transition occurs.

**A** Using equation (3.10), and the value for $R_H$ calculated in Worked Problem 3.2:

(a) $\tilde{\nu} = 1.096\ 79 \times 10^5\ \text{cm}^{-1}\ (1/1^2 - 1/5^2)$
$= 105\ 292\ \text{cm}^{-1}$
$\therefore\ \lambda = 1/105\ 292 \times 100\ \text{m}^{-1}$
$= 94.9741$ nm, in the far-ultraviolet region
(b) $\tilde{\nu} = 1.096\ 79 \times 10^5\ \text{cm}^{-1}\ (1/3^2 - 1/20^2)$
$= 11\ 912.4\ \text{cm}^{-1}$, in the near-infrared region
(c) $\tilde{\nu} = 1.096\ 79 \times 10^5\ \text{cm}^{-1}\ (1/102^2 - 1/154^2)$
$= 5.917\ 27\ \text{cm}^{-1}$, in the millimetre wave region
(d) $\tilde{\nu} = 1.096\ 79 \times 10^5\ \text{cm}^{-1}\ (1/153^2 - 1/154^2)$
$= 6.065\ 06 \times 10^{-2}\ \text{cm}^{-1}$
Multiply by the speed of light to convert to frequency:
$\therefore\ \nu = 2.997\ 92 \times 10^{10}\ \text{cm s}^{-1} \times 6.065\ 06 \times 10^{-2}\ \text{cm}^{-1}$
$= 1.818\ 26 \times 10^9\ \text{s}^{-1}$
$= 1.818\ 26$ GHz, in the microwave region

An important extension of the theory of the hydrogen atom, going far beyond the fitting of the Balmer series in equation (3.2), was due to Bohr. The Bohr theory regarded the electron classically as a particle travelling in circular orbits around the nucleus, each orbit characterized by a value of the quantum number $n$. The fact that it can only be in certain orbits is due to its angular momentum, $p_\theta$, where $\theta$ is the angle of rotation around the nucleus, taking only certain values given by:

$$p_\theta = nh/2\pi \qquad (3.13)$$

The discovery of the **photoelectric effect** in 1887 resulted in further evidence for the quantization of energy. The original experiment was performed on the surface of solid sodium, *in vacuo*, which was subjected to radiation of varying wavelength from the visible to the near ultraviolet.

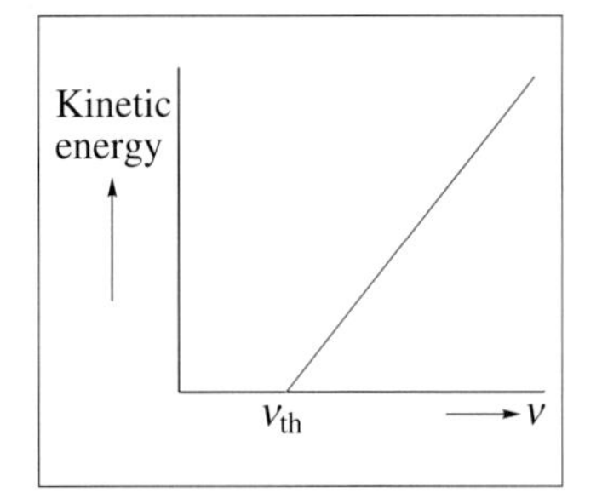


**Figure 3.4** Graphical illustration of the photoelectric effect

As the frequency of the radiation was increased, nothing happened until the **threshold frequency**, $\nu_{th}$, was reached. Beyond the threshold frequency, photoelectrons are emitted from the surface with kinetic energy which increases linearly with the frequency, as shown in Figure 3.4. This observation was clear evidence of the quantization of the radiation falling on the surface, each incident photon having an energy $h\nu$. Above the threshold frequency, this energy is sufficient to eject a photoelectron from the surface and provide it with kinetic energy $\frac{1}{2}m_e v^2$, where:

$$h\nu = I + \tfrac{1}{2}m_e v^2 \tag{3.14}$$

In some older texts, the ionization energy is referred to as the ionization potential.

and $m_e$ and $v$ are the mass and velocity, respectively, of the photoelectron. $I$ is the ionization energy of a sodium atom at the surface of the metal. Although the SI unit of ionization energy is the joule, the unit of electronvolt is commonly used (see equations 2.12 and 2.13).

### Worked Problem 3.4

**Q** The ionization energy of a free sodium atom is 5.139 eV. Calculate the kinetic energy of a resulting photoelectron when a sodium atom in the gas phase is irradiated with near-ultraviolet radiation of wavelength 230.5 nm.

**A** From equation (2.12):

$5.139 \text{ eV} = 5.139 \times 1.602 \times 10^{-19} \text{ J} = 8.233 \times 10^{-19} \text{ J}$

From equations (2.4) and (2.9), the energy of a photon of wavelength 230.5 nm is given by:

$E = hc/\lambda = 6.626 \times 10^{-34} \text{ J s} \times 2.998 \times 10^{8} \text{ m s}^{-1}/230.5 \times 10^{-9} \text{ m} = 8.618 \times 10^{-19} \text{ J}$

$\therefore$ kinetic energy of the photoelectron $= (8.618 - 8.233) \times 10^{-19} \text{ J} = 3.85 \times 10^{-20} \text{ J}$

The ionization energy of a free sodium atom is much greater than that of an atom on the surface of solid sodium. As a result, it requires far ultraviolet radiation to ionize a free sodium atom, which contrasts with the visible radiation which is sufficient to ionize metallic sodium.

In 1900, Planck rationalized experimental observations, such as the spectrum of the hydrogen atom and the photoelectric effect, by the introduction of quantum theory, the quantization of energy, and the Planck constant, $h$.

A further important step in the development of theory in relation to spectroscopy was the introduction of the **de Broglie equation**, in 1924. Not only can all electromagnetic radiation be regarded as consisting of either waves or particles, but very light "particles" can also behave as waves. One of the most important examples is the electron. Although we often think of the electron as a particle, the phenomenon of **electron diffraction** shows that it can also behave as a wave.

The first experiment to demonstrate diffraction of electrons was carried out by Davisson and Germer in 1925. They showed that a monochromatic electron beam is reflected and diffracted by a surface of crystalline nickel. The electrons in the beam are diffracted by the metal atoms on the surface, rather as visible light is diffracted on passing through a narrow slit. A beam of electrons passing through an extremely thin metal foil is diffracted in a similar way.

The de Broglie equation reconciles, in a remarkably simple way, this dual particle–wave nature as follows:

$$p = h/\lambda \tag{3.15}$$

where $\lambda$ is the wavelength of the wave and $p$ the momentum of the particle. For example, for an electron travelling in a straight line with velocity $v$:

$$p = m_e v \tag{3.16}$$

where $m_e$ is the mass of the electron.

The concept of a particle, such as an electron, acting as both a particle and a wave gave rise to the **uncertainty principle**, formulated by Heisenberg in 1927. The principle states that, if the momentum, $p_x$, or wavelength, of a particle moving in the $x$-direction is known exactly, the position, $x$, is completely uncertain. Conversely, if the position is known exactly, the momentum is completely uncertain. The uncertainties, $\Delta p_x$ and $\Delta x$, are related by:

$$\Delta p_x \Delta x \geq h/2\pi \tag{3.17}$$

**Worked Problem 3.5**

**Q** Calculate the velocity of an electron which, treated as a wave, has a wavelength of 196 nm.

**A** From equations (3.15) and (3.16):

$$v = h/\lambda m_e = 6.626 \times 10^{-34}\ \text{J s}/196 \times 10^{-9}\ \text{m} \times 9.109 \times 10^{-31}\ \text{kg} = 3710\ \text{m s}^{-1}$$

Because all the units used in the calculation have been expressed in SI base units, the units of the calculated velocity must also be base units, *i.e.* m $s^{-1}$. If you wish to double-check this, use the information that the base units of J are $m^2$ kg $s^{-2}$.

The discovery of the dual wave–particle nature of the electron prompted a rethink of the classical Bohr picture in which the electron in the hydrogen atom behaves as a particle moving in circular orbits, rather like a planet orbiting a star. In fact, the wave picture of the electron explains beautifully the restricted nature of these orbits. The electron can be only in orbits in which the wave is a **standing wave**. This means that the orbit, of circumference $2\pi r$, where $r$ is the radius of the orbit, must contain an integral number of wavelengths $\lambda$:

$$n\lambda = 2\pi r \tag{3.18}$$

The integer $n$, the **principal quantum number**, can take the values $n = 1, 2, 3, 4, \ldots \infty$, and is the quantum number introduced by Bohr.

In the wave picture of the electron, the orbits are referred to as **orbitals**. Figure 3.5 illustrates the standing wave in the orbital with $n = 12$.

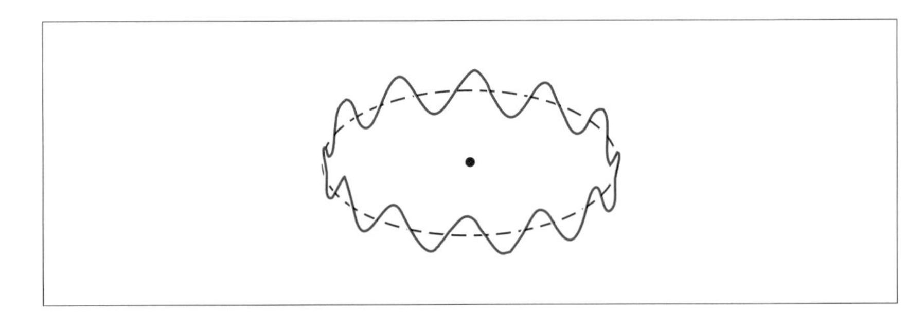

**Figure 3.5** The standing wave in an atomic orbital with $n = 12$

## 3.3 Results of Applying Quantum Mechanics to the Hydrogen Atom

The behaviour of an electron as a wave rather than as a particle requires a quite different theoretical approach to that of, for example, Bohr as applied to the hydrogen atom. This approach is that of **quantum mechanics**, sometimes referred to as **wave mechanics**.

The basis of the subject of quantum mechanics is the **Schrödinger equation**, which can be written in the deceptively simple form:

$$H\psi = E\psi \tag{3.19}$$

in which $E$ represents the discrete energy levels of the system, atom or molecule which is under consideration. These values of $E$ are also called **eigenvalues**. The quantity $\psi$ represents the **wave function**, or **eigenfunction**, corresponding to each eigenvalue. This function describes the form of the wave for each energy level.

The Schrödinger equation may be used to obtain the wave functions corresponding to the energy levels which are known from spectroscopic measurements. On the other hand, assumed forms of the wave functions may be used to predict the energy levels of an atom or molecule. The close relationship between quantum mechanics and spectroscopy is a consequence of this interrelation between wave functions and energy levels.

The simplicity of equation (3.19) is misleading: $H$, the **hamiltonian**, is the sum of the **potential energy**, $-e^2/4\pi\varepsilon_0 r$, due to coulombic attraction between the electron and nucleus a distance $r$ apart, and the *quantum mechanical equivalent* of the **kinetic energy**:

$$H = -(e^2/4\pi\varepsilon_0 r) - (\hbar^2/2\mu)\nabla^2 \qquad (3.20)$$

where $\mu$ is the reduced mass of the electron and proton, given in equation (3.5). The significance of $\nabla^2$, and the solution of the Schrödinger equation for $E$ and $\psi$, are described in books listed under "Further Reading".

Some of the most important wave functions resulting from solution of the Schrödinger equation are illustrated in Figure 3.6. This figure shows the wave functions for the so-called 1s, $2p_x$, $2p_y$, $2p_z$, $3d_{z^2}$, $3d_{xz}$, $3d_{yz}$, $3d_{x^2-y^2}$ and $3d_{xy}$ atomic orbitals. The wave functions are drawn in the form of boundary surfaces, within which there is a high probability (say 90%) of finding the electron. These wave functions appear very different from that in, for example, Figure 3.5 in which the electron was confined to moving in a circle rather than in three-dimensional space.

The labels for the orbitals indicate the relevant values of various quantum numbers. The 1, 2 or 3 label is the value of the principal quantum number $n$, as in the Bohr theory of the hydrogen atom. The s, p or d label indicates the value of the **azimuthal quantum number** $\ell$ where:

$$\ell = 0, 1, 2, 3, \ldots (n-1) \qquad (3.21)$$

and therefore can take $n$ values. This quantum number describes the discrete values that the **orbital angular momentum** of the electron can take. The association between the s, p, d, ... labels and the values of $\ell$ is as follows:

$$\begin{array}{lcccccc} \ell = & 0 & 1 & 2 & 3 & 4 & 5 \\ & \text{s} & \text{p} & \text{d} & \text{f} & \text{g} & \text{h} \end{array} \qquad (3.22)$$

There is a further quantum number, $m_\ell$, associated with the orbital angular momentum and known as the **magnetic quantum number**. It can take $2\ell + 1$ values, given by:

$$m_\ell = 0, \pm1, \pm2, \pm3, \ldots \pm\ell \qquad (3.23)$$

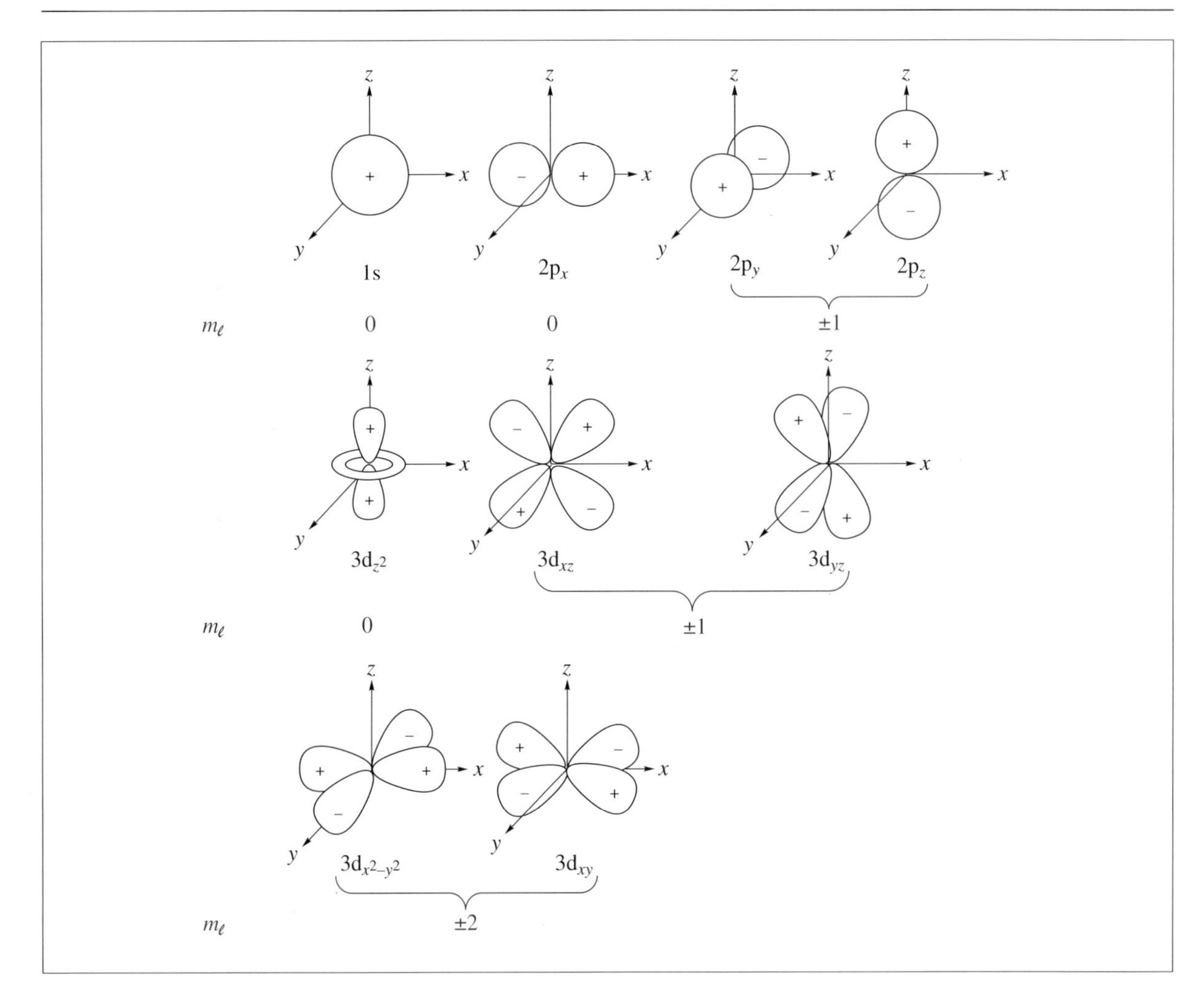


**Figure 3.6** Wave functions for various atomic orbitals of the hydrogen atom

The orbital angular momentum of the electron is a **vector** quantity: it has not only magnitude but also direction. When the atom is in a magnetic field, this vector can lie only in certain directions in relation to the field. For example, for $\ell = 3$, $m_\ell$ can take the values 3, 2, 1, 0, –1, –2, –3. The directions that the angular momentum vector can take for a magnetic field in the $z$-direction, corresponding to the seven values of $m_\ell$, are illustrated in Figure 3.7.

The 2p and 3d orbitals illustrated in Figure 3.6, in which the $z$-axis is taken, arbitrarily, to be the unique axis, are distinguished by subscripts which refer to the values of $m_\ell$ as indicated in the figure. The $2p_z$ and $3d_{z^2}$ orbitals correspond uniquely to $m_\ell = 0$, but those involving the $x$- and/or $y$-axes do not correspond to unique values of $m_\ell$. Although it is conventional to draw the 3d orbitals as if the $z$-axis is unique, the spherically symmetrical nature of the atom dictates that either the $x$- or $y$-axis could be taken to be unique.

In the absence of a magnetic (or electric) field, all the 2p orbitals are **degenerate**. This means that they have the same energy but different wave functions. Similarly, all the 3d orbitals are degenerate. Because there are three 2p orbitals and five 3d orbitals, they are said to be three-fold and five-fold degenerate, respectively.

We shall encounter degeneracy in several contexts throughout this book. The fact that an atom or molecule has states which are degenerate always means that the states have the same energy but are described by different wave functions.

The hydrogen atom (and all one-electron atoms, such as $He^+$ and $Li^{2+}$) is unusual in two respects: one is that the Schrödinger equation can be solved exactly for it, and the other is that all orbitals with the same value of *n*, for example 2s, $2p_x$, $2p_y$ and $2p_z$, are degenerate, as Figure 3.3 indicates. For all other atoms, even for He which has only two electrons, this is not the case.

There are two further angular momenta in the hydrogen atom which give rise to quantized energy. These angular momenta are due to **electron spin** and **nuclear spin**. In the particle picture of the electron, the concept of it spinning on its own axis is a valid one, but this is not the case in the wave picture. Nevertheless, there is an electron spin quantum number *s* associated with the electron, and it can take only one value, $\frac{1}{2}$. This applies also to each electron in all atoms. The quantum number *I*, associated with nuclear spin, takes the value $\frac{1}{2}$ for the hydrogen atom ($^1H$) but its value depends on the particular nucleus. For example, for $^2H$, $I = 1$, for $^{12}C$, $I = 0$, and for $^{13}C$, $I = \frac{1}{2}$.

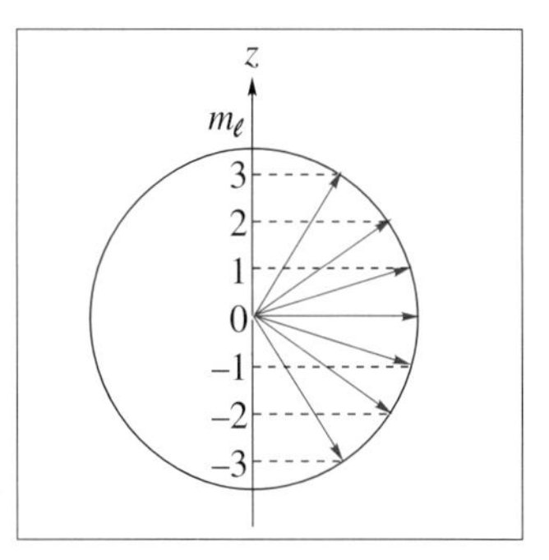


**Figure 3.7** In a magnetic field, the angular momentum vector, for $\ell = 3$, can take seven different orientations

In a magnetic field the electron spin and nuclear spin angular momentum vectors may take up only certain orientations with respect to the field, in a similar way to the orbital angular momentum vector, as shown in Figure 3.7. These directions are determined by the values that the quantum numbers $m_s$ and $m_I$ can take, where:

$$m_s = s, s-1, \ldots -s = +\tfrac{1}{2} \text{ and } -\tfrac{1}{2}, \text{ for one electron} \qquad (3.24)$$

and

$$m_I = I, I-1, \ldots -I = +\tfrac{1}{2} \text{ and } -\tfrac{1}{2}, \text{ for the } ^1\text{H and } ^{13}\text{C nuclei and } 1, 0 \text{ and } -1, \text{ for the } ^2\text{H nucleus} \qquad (3.25)$$

The effects of nuclear spin are of great importance in nuclear magnetic resonance (NMR) spectroscopy; in atomic spectra, the effects are extremely small and will not concern us further.

## Summary of Key Points

**1.** *Meaning of quantization*
In atoms and molecules, energy levels are quantized, which means that they are precise or discrete.

**2.** *Quantization in the hydrogen atom*
The spectrum of the hydrogen atom shows an infinite series of lines, and each series converges as the energy increases. Theoretical treatment by Balmer and Bohr.

**3.** *Photoelectric effect*
Definition of threshold frequency and ionization energy.

**4.** *Dual wave–particle nature of the electron*
Electron diffraction and the de Broglie equation.

**5.** *Quantum mechanics and the hydrogen atom*
The Schrödinger equation, wave functions and energy levels. Principal, azimuthal, magnetic, electron spin and nuclear spin quantum numbers. s, p, d, f, ... atomic orbitals. Degeneracy.

## Problems

**3.1.** Calculate, to four significant figures, the angular momentum of the electron in the hydrogen atom when the electron is in the (a) $n = 5$ and (b) $n = 100$ orbital.

**3.2.** Calculate the wavelength of a proton travelling in a straight line with a velocity of 6034 m $s^{-1}$.

**3.3.** In the absence of a magnetic or electric field, what is the degree of degeneracy of (a) f orbitals and (b) h orbitals in the hydrogen atom?

**3.4.** Calculate, to six significant figures, the wavenumber of the first line ($n' = 2$, $n'' = 1$) of the Lyman series for $^{6}Li^{2+}$.

# 4
# Quantization in Polyelectronic Atoms

**Aims**

In this chapter you will be introduced to:

- The effects of electron repulsion in polyelectronic atoms
- The aufbau and Pauli exclusion principles, which describe the feeding of electrons into orbitals
- A picture of the Periodic Table which results from applying these principles
- Spin multiplicity, singlet and triplet states and selection rules in the helium atom
- Coupling of orbital and spin angular momenta in polyelectronic atoms
- The special cases of equivalent electrons
- Russell–Saunders coupling and Hund's rules
- Splitting of spin mutliplets; Landé interval rule; normal and inverted multiplets
- Selection rules for emission and absorption spectra

## 4.1 Effects of More than One Electron in an Atom

The comparative simplicity of the spectrum of the hydrogen atom, and one-electron ions, and of the theoretical interpretation is lost as soon as a second electron is present. The Schrödinger equation is no longer exactly soluble, which means that the electronic energy levels are not exactly calculable: approximations must be made. So far as the experimentally observed spectrum is concerned, perhaps the most important difference from a one-electron atom is that the $(2\ell + 1)$-fold degeneracy of the s, p, d, ... orbitals is removed. This is shown by the typical set of orbital energies illustrated in Figure 4.1.

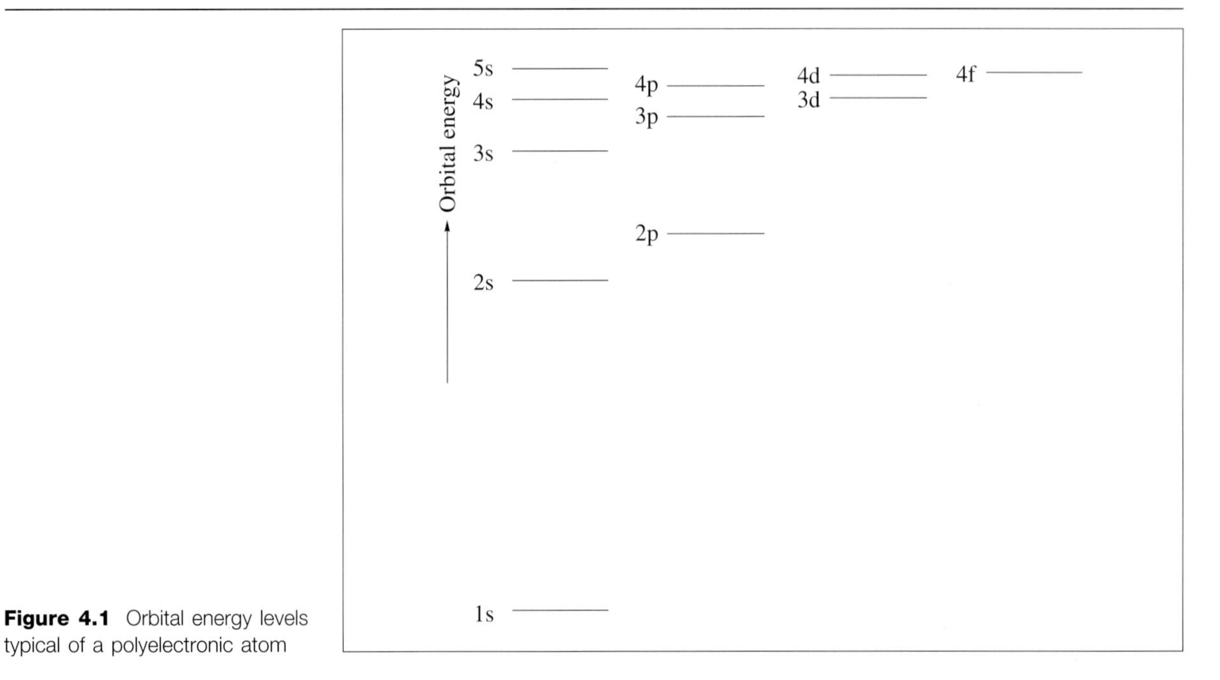


**Figure 4.1** Orbital energy levels typical of a polyelectronic atom

When there is just one electron in the atom there is only one **coulombic force** present, the force of **attraction** between the positively charged nucleus and the negatively charged electron. The introduction of a second electron, as in the helium atom, or more electrons, as in any other polyelectronic atom, adds a second type of coulombic force, that of **repulsion** between the negatively charged electrons. One very important result of this is that, unlike the case of one-electron atoms, orbital energies and wave functions can no longer be calculated exactly.

The orbital energies illustrated in Figure 4.1, up to the 5s orbital, are not quantitative and, indeed, are different for different atoms. However, a useful simplifying factor is that the order of the orbital energies is maintained for most atoms.

For polyelectronic atoms in their lowest energy state (the ground state), the available electrons are fed into the available orbitals, in order of increasing energy, according to the **aufbau principle** or building-up principle. In addition, the **Pauli exclusion principle** must be adhered to. This principle states that no two electrons may have the same set of quantum numbers $n$, $\ell$, $m_\ell$ and $m_s$. Therefore the 1s orbital, and any other s orbital, can accommodate only two electrons for which $\ell = 0$, $m_\ell = 0$, and $m_s = \pm\frac{1}{2}$. The two electrons are said to have their spins paired or to have anti-parallel spins. Similarly, any p orbital can accommodate six electrons for which $\ell = 1$, $m_\ell = 0, \pm1$, and $m_s = \pm\frac{1}{2}$, and any d orbital can accommodate ten electrons for which $\ell = 2$, $m_\ell = 0, \pm1, \pm2$, and $m_s = \pm\frac{1}{2}$.

### Worked Problem 4.1

**Q** Show that an f orbital can accommodate 14 electrons.

**A** For an f orbital, for example 4f:

$\ell = 3$, $m_\ell = +3, +2, +1, 0, -1, -2,$ or $-3$, and $m_s = +\frac{1}{2}$ or $-\frac{1}{2}$

Therefore there are 14 different combinations of this set of quantum numbers that each of the 14 electrons may have.

Using the aufbau and Pauli principles, a very useful picture of the **Periodic Table** of the elements may be built up by feeding all the available electrons into the orbitals to give the **ground configuration** (the lowest energy electron configuration, corresponding to the ground state) for each element. Examples of these are:

He $1s^2$ or K
Li $1s^22s^1$ or $K2s^1$
or $[He]2s^1$
B $1s^22s^22p^1$ or $K2s^22p^1$
or $[He]2s^22p^1$
Ne $1s^22s^22p^6$ or KL
Na $1s^22p^63s^1$ or $KL3s^1$
or $[Ne]3s^1$
Mg $1s^22s^22p^63s^2$ or $KL3s^2$
or $[Ne]3s^2$
K $1s^22p^63s^23p^64s^1$ or $KL3s^23p^64s^1$
or $[Ar]4s^1$
Ti $1s^22s^22p^63s^23p^63d^24s^2$ or $KL3s^23p^63d^24s^2$
or $[Ar]3d^24s^2$
Cr $1s^22s^22p^63s^23p^63d^54s^1$ or $KL3s^23p^63d^54s^1$
or $[Ar]3d^54s^1$
Cu $1s^22s^22p^63s^23p^63d^{10}4s^1$ or $KLM4s^1$
or $[Ar]3d^{10}4s^1$
Zr $1s^22s^22p^63s^23p^63d^{10}4s^24p^64d^25s^2$ or $KLM4s^24p^64d^25s^2$
or $[Kr]4d^25s^2$
Nd $1s^22s^22p^63s^23p^63d^{10}4s^24p^64d^{10}4f^45s^25p^66s^2$
or $KLM4s^24p^64d^{10}4f^45s^25p^66s^2$
or $[Xe]6s^2$

The post-superscript attached to an orbital symbol indicates the number of electrons in that orbital.

From these examples some important points emerge:

1. Orbitals with the same value of the principal quantum number $n$ comprise a **shell**.
2. Sometimes, but mostly in older texts, filled shells with $n = 1, 2, 3, 4, \ldots$ are abbreviated to K, L, M, N, ... This leads to somewhat abbreviated configuration labels.
3. Configurations can be abbreviated more fully by indicating the configuration of a noble gas (He, Ne, Ar, Kr, Xe ...) by [He], [Ne], [Ar], [Kr], [Xe], ...
4. Orbitals with the same values of $n$ and $\ell$, for example the 1s, 2p, 3p and 3d orbitals, are referred to as **sub-shells**.
5. The **noble gases**, He, Ne, Ar, ..., have completely filled p sub-shells or, in the case of helium, a filled 1s shell, and these configurations confer extreme unreactivity on these elements.
6. The **alkali metal** atoms, Li, Na, K, ..., are characterized by a single electron in an outer s orbital. This configuration explains the ability of these elements to form monopositive ions (*i.e.* to be monovalent).
7. The **alkaline earth metal** atoms, Be, Mg, Ca, ..., are characterized by two electrons in an outer s orbital. This explains their ability to form dipositive ions (*i.e.* to be divalent).
8. The **transition metal** atoms are characterized by electrons in incomplete d subshells. The 3d sub-shell can accommodate 10 electrons. As a result, there are 10 members of the first transition series of elements, Sc, Ti, V, ... However, the first member, Sc, has only one 3d electron and does not possess the properties typical of a transition metal. Figure 4.1 shows that the 3d and 4s orbitals are very similar in energy so that, in some transition metal elements, it is touch-and-go whether the 4s orbital is filled first. In Ti, with an outer $3d^24s^2$ configuration, the 4s orbital is filled first, but in Cr, with an outer $3d^54s^1$ configuration, it is favourable for the 3d sub-shell to be half-filled in preference to filling the 4s orbital. This innate stability associated with a half-filled sub-shell is typical and is a result of the electrons preferring to be distributed with one in each of the 3d orbitals with parallel spins. There is also a greater stability associated with a filled 3d sub-shell than with a filled 4s shell. As a result, the outer configuration for the ground state of Cu is $3d^{10}4s^1$.
9. In the second transition series, Y, Zr, Nb, ..., the 4d sub-shell is partially occupied.
10. In the **lanthanide** elements, Ce, Pr, Nd, ..., the 4f sub-shell is partially occupied, while in the **actinide** elements, Pa, U, Np, ..., the 5f sub-shell is partially occupied.

**Worked Problem 4.2**

**Q** What are the ground configurations of the (a) Ar, (b) Fe and (c) Mo atoms?

**A** (a) Ar: $1s^2 2s^2 3s^2 3p^6$. Typical of a noble gas, argon has an outer, filled p sub-shell.
(b) Fe: [Ar]$3d^6 4s^2$. A member of the first transition series, iron has a partially filled 3d sub-shell.
(c) Mo: [Kr]$4d^5 5s^1$. A member of the second transition series, molybdenum has a partially filled 4d sub-shell. Similar to chromium in the first transition series, molybdenum prefers to have a half-filled 4d sub-shell, conferring stability, at the expense of there being only one electron in the 5s orbital.

## 4.2 The Helium Atom

In many ways the helium atom behaves as a prototype for the spectroscopic behaviour of other polyelectronic atoms, and even of some molecules, particularly those with ground electronic states which arise from **closed shell** configurations, that is configurations in which all the electrons are in filled orbitals.

Helium, in its ground configuration $1s^2$, is the simplest closed shell atom. There is an important rule that, for all closed shells, the result of the addition of the angular and spin momenta of all the electrons in the shell is that:

1. There is no net orbital angular momentum, *i.e.* $L = 0$, where the azimuthal quantum number $\ell$ for a one-electron atom (see equation 3.21) is replaced by $L$ for a polyelectronic atom.
2. There is no net electron spin angular momentum, *i.e.* $S = 0$, where the quantum number $s$ for a one-electron atom is replaced by $S$ for a polyelectronic atom.

There is only one **term**, that with $L = 0$ and $S = 0$, arising from this ground configuration. The conventional label for this term is $^1S$. The pre-superscript "1" is the value of the **multiplicity** where:

$$\text{electron spin multiplicity} = 2S + 1 \qquad (4.1)$$

This is the number of values that $M_S$ can take, and these values are:

$$M_S = S, S - 1, \ldots -S \qquad (4.2)$$

What we have called a **term** here, such as the $^1S$ term, is often referred to as a state, as in the $^1S$ state. However, this is not to be recommended. We retain the word "state" to describe what often comprises a component of a term. For example, we shall see in Section 4.3.4 that a term such as $^3P$ has three components, labelled $^3P_2$, $^3P_1$ and $^3P_0$, which have slightly different energies. It is states such as these which are involved in spectroscopic observations and therefore are of much greater experimental importance than terms which have only theoretical significance.

The number of values that $M_S$ can take is also the number of components into which the term might be split in a magnetic field. In this case the multiplicity is 1, and the $^1$S term is said to be a **singlet** term.

Analogous to the labelling of s, p, d, f, ... orbitals, indicated in equation (3.22), states of polyelectronic atoms are labelled according to the value of $L$ as follows:

$$\begin{array}{lccccc} L = & 0 & 1 & 2 & 3 & 4\ldots \\ & \mathrm{S} & \mathrm{P} & \mathrm{D} & \mathrm{F} & \mathrm{G}\ldots \end{array} \tag{4.3}$$

Consequently, the ground term of helium is an S term.

Excited terms of helium result from the **promotion** of one of the electrons from the 1s to a higher energy orbital. The probability of both 1s electrons being promoted is sufficiently low, and requires such a large amount of energy, that such excited terms will not concern us here.

When one electron is promoted, the spin quantum number $m_s$ $(= \pm\frac{1}{2})$ may be unchanged or change sign. Then the electrons are said to have **antiparallel** or **parallel** spins, respectively. When they are antiparallel, the two spins cancel, $S = \frac{1}{2} - \frac{1}{2} = 0$, and:

$$\text{electron spin multiplicity} = 2S + 1 = 1 \tag{4.4}$$

Therefore the term is, like the ground term, a singlet term. When the electron spins are parallel, $S = \frac{1}{2} + \frac{1}{2} = 1$, the multiplicity is 3 and the resulting term is a **triplet** term. Such singlet and triplet terms are illustrated in Figure 4.2. The case in Figure 4.2(d), in which both electrons are in the same orbital with parallel spins, is forbidden by the Pauli principle as they would both have the same set of quantum numbers.

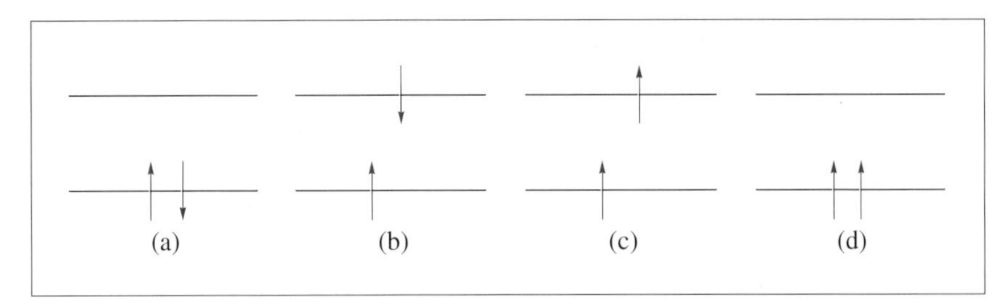


**Figure 4.2** Possible distribution of electron spins between two orbitals to give (a) a ground singlet term, (b) an excited singlet term, (c) an allowed excited triplet term and (d) a forbidden triplet term

We can see now that all the electronic terms of the hydrogen atom have the same multiplicity. Since $S$ can only be $\frac{1}{2}$, $2S + 1 = 2$, and they are all **doublet** terms. However, in the case of helium, there are two sets of terms, a **manifold** of singlet terms, including the ground term, and a manifold of triplet excited terms. Some of the lower energy terms in these manifolds are illustrated in Figure 4.3. The labelling of the terms is by the multiplicity, 1 or 3, as a pre-superscript and by S, P, D, ... to indicate the value of $L$: for excited terms, this is simply the value of $\ell$ for the promoted electron, since $\ell = 0$ for the electron remaining in the 1s orbital. The terms included in Figure 4.3 are of the type $^1S$, $^1P$, $^1D$, $^3S$, $^3P$ and $^3D$.

For the hydrogen atom we saw that the only selection rule, governing transitions between terms, that we required was that $\Delta n$ is unrestricted. For transitions in the helium atom the corresponding selection rules are:

$$\Delta n \text{ is unrestricted, } \Delta\ell = \pm 1,\ \Delta S = 0 \tag{4.5}$$

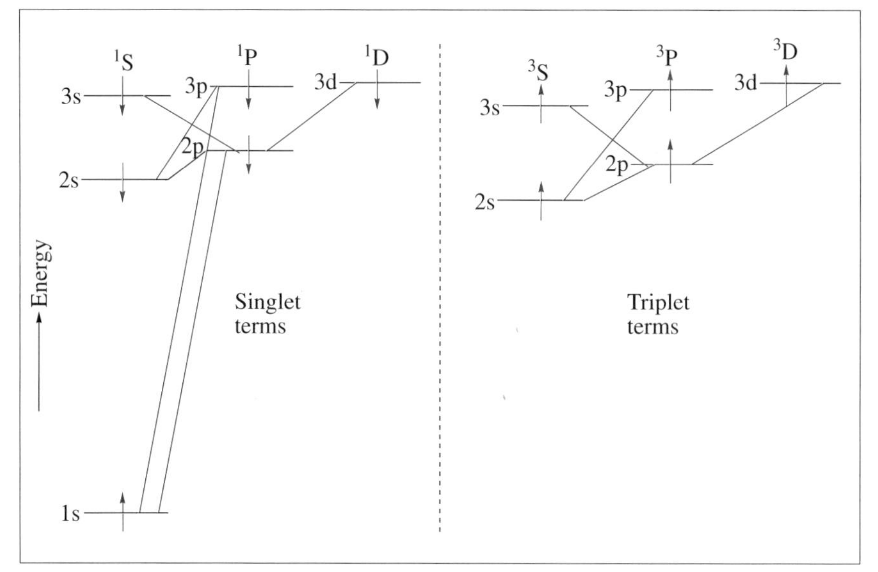


**Figure 4.3** Energy level diagram showing singlet and triplet terms of the helium atom

The $\Delta\ell$ selection rule refers to the change of orbital angular momentum for the promoted electron, assuming that the other one remains in the 1s orbital. The result of the $\Delta S$ selection rule is that no transitions are allowed between singlet and triplet terms. Some of the lower energy allowed transitions are indicated in Figure 4.3.

Typically, for terms of any atom which differ only in their multiplicity, the term of highest multiplicity lies lowest in energy. In the case of helium, Figure 4.3 shows that each triplet term lies below the corresponding singlet term.

Figure 4.3 also shows that no transitions occur between the triplet and singlet term manifolds. As a result of, say, an electrical discharge in helium gas the atoms will be excited into a wide range of singlet or triplet terms. Many of them will arrive in either of the lowest energy excited terms, $2s^1S$ or $2s^3S$, as a result of emission or collision processes. The transition from the $2s^1S$ to the ground term is forbidden by the $\Delta\ell$ selection rule. The $2s^1S$ atom is therefore unusually long-lived and is said to be **metastable**. The transition from the $2s^3S$ term is doubly forbidden, by the $\Delta\ell$ and $\Delta S$ selection rules, and the $2s^3S$ atom is also metastable with an even longer lifetime, of about 1 ms, in a typical discharge.

Figure 4.3 shows only the lower energy transitions which are found in the emission spectrum of helium, but they form series of lines, each series converging at high wavenumber to the ionization limit. This smoothly convergent behaviour is reminiscent of the spectrum of hydrogen but, because of the problems associated with repulsion between the two electrons, their wavenumbers are much more difficult to obtain theoretically.

The wavenumbers of possible transitions between singlet and triplet terms can be predicted very accurately from the known singlet and triplet energy levels. However, very careful search for such transitions has been unsuccessful: the $\Delta S = 0$ selection rule is very rigidly obeyed in the helium atom.

## 4.3 Other Polyelectronic Atoms

### 4.3.1 Coupling of Orbital Angular Momenta

In the examples of the hydrogen and helium atoms, in either the ground or any of the excited electronic terms that we have encountered, there has been, at most, only one electron with non-zero angular momentum, *i.e.* with $\ell \neq 0$, for example the $1s^12p^1$ configuration of helium. In some doubly excited configurations of helium, such as $2p^13d^1$, both electrons may have $\ell \neq 0$, but these are of such high energy that they have not concerned us.

In other polyelectronic atoms, however, there is a possibility of there being two or more electrons with $\ell \neq 0$ in a much lower energy configuration. Such an example is the excited configuration:

Each type of angular momentum, associated with the orbital or spin motion of a charged particle, such as an electron or nucleus, creates a **magnetic moment**. This acts rather like a tiny bar magnet. In this way, we can think of all the various angular momenta, resulting from the electron orbital motion, electron spin and nuclear spin, as being represented by a multitude of bar magnets. In principle, all these bar magnets can interact, or couple, with each other in a very complex manner. Fortunately, some of the interactions are very much stronger than others, resulting in considerable simplification of the way we can treat them theoretically.

$$\text{C} \quad 1s^22s^22p^13d^1 \tag{4.6}$$

of the carbon atom. In this case, the 2p electron has $\ell_1 = 1$ and the 3d electron has $\ell_2 = 2$. The 1s and 2s electrons all have $\ell = 0$, and need not be considered further.

In such a case, the two orbital angular momenta are **coupled** to give the **total orbital angular momentum**. This is also quantized, and the quantum number associated with it is $L$ which can take the values:

$$L = \ell_1 + \ell_2, \ell_1 + \ell_2 - 1, \ldots |\ell_1 - \ell_2| \tag{4.7}$$

where $|\ell_1 - \ell_2|$ is the **modulus** of $\ell_1 - \ell_2$. The quantity $|\ell_2 - \ell_1|$ is defined as $\ell_1 - \ell_2$ or $\ell_2 - \ell_1$, whichever is the positive quantity; in this case, it is $\ell_2 - \ell_1$. Therefore for the configuration in equation (4.6), $L = 3$, 2 and 1, and, according to equation (4.3), F, D and P terms result.

## Box 4.1 Vector Representation of Orbital Angular Momentum

The orbital angular momentum of an electron is a vector quantity, which means that it has magnitude and direction. The magnitude is given by:

$$[\ell(\ell + 1)]^{\frac{1}{2}}\hbar = \ell^*\hbar \tag{4.8}$$

where $\ell^*$ is a convenient abbreviation for $[\ell(\ell + 1)]^{\frac{1}{2}}$.

Vector diagrams can be drawn to illustrate the case represented by equation (4.7). The vectors representing the orbital angular momenta $\ell_1$ and $\ell_2$, of magnitudes $[\ell_1(\ell_1 + 1)]^{\frac{1}{2}}\hbar$ and $[\ell_2(\ell_2 + 1)]^{\frac{1}{2}}\hbar$, can take up only certain orientations with respect to each other, such that the resultant vector is of magnitude:

$$[L(L + 1)]^{\frac{1}{2}}\hbar = L^*\hbar \tag{4.9}$$

where $L$ is given by equation (4.7). In this case it is of magnitude $12^{\frac{1}{2}}\hbar$, $6^{\frac{1}{2}}\hbar$ or $2^{\frac{1}{2}}\hbar$, corresponding to $L = 3$, 2 or 1, respectively. These vector diagrams are shown in Figure 4.4.

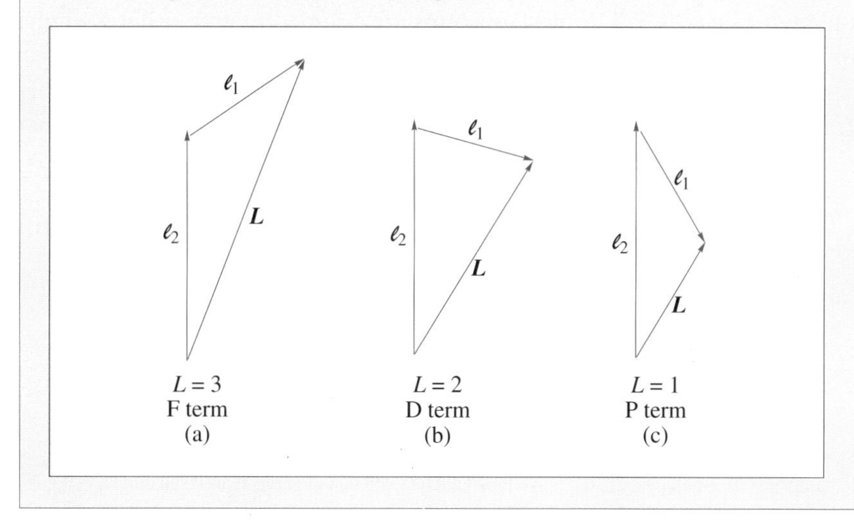


**Figure 4.4** Vector diagrams for the coupling of the orbital angular momenta of a p and a d electron to give (a) an F, (b) a D and (c) a P term

Although this treatment tells us which terms arise from a particular configuration, it does not tell us anything about the relative energies of these terms. It would require a quantum mechanical calculation to do this.

**Worked Problem 4.3**

**Q** What terms result from (a) the $1s^22s^22p^14f^1$ and (b) the $1s^22s^23d^14f^1$ excited configurations of the carbon atom?

**A** (a) The 1s and 2s electrons contribute nothing to $L$. For the 2p and 4f electrons, $\ell_1 = 1$ and $\ell_2 = 3$, respectively. Therefore from equation (4.7), $L = 1 + 3, 1 + 3 - 1$ and $3 - 1 = 4$, 3 and 2, giving G, F and D terms.
(b) In this case, $L = 2 + 3, 2 + 3 - 1, \ldots 3 - 2 = 5$, 4, 3, 2 and 1, giving H, G, F, D and P terms. This doubly excited configuration, in which both 2p electrons have been promoted, gives rise to excited terms of very high energy.

### 4.3.2 Coupling of Electron Spin Angular Momenta

We have seen in Section 4.2, for the simple case of the two-electron helium atom, that the excited terms may be singlet or triplet, depending on whether the two electron spins are antiparallel, when $S = \frac{1}{2} - \frac{1}{2} = 0$ and the multiplicity $(2S + 1)$ is 1, or parallel, when $S = \frac{1}{2} + \frac{1}{2} = 1$ and the multiplicity is 3. The electron spins are coupled to give either a singlet or a triplet excited term.

Analogous to the total orbital angular momentum quantum number $L$ in equation (4.7), the **total spin angular momentum** quantum number $S$ can, in general, take the values:

$$S = s_1 + s_2, s_1 + s_2 - 1, \ldots |s_1 - s_2| \tag{4.10}$$

which, in the case of two electrons, can be only 1 or 0.

**Box 4.2 Vector Representation of Electron Spin Angular Momentum**

Electron spin angular momenta are also vector quantities, having magnitude and direction. The magnitude of the vector is given by:

$$[s(s + 1)]^{\frac{1}{2}}\hbar = s^*\hbar \tag{4.11}$$

where $s^* = [s(s + 1)]^{\frac{1}{2}}$. This is analogous to the magnitude of the orbital angular momentum in equation (4.8), except that $s$ can only be $\frac{1}{2}$ and therefore the magnitude can only be $(3^{\frac{1}{2}}/2)\hbar$.

Vectors, for the case of two electrons with $s_1 = s_2 = \frac{1}{2}$, are shown in Figure 4.5. The magnitude of the resultant vector is given by:

$$[S(S + 1)]^{\frac{1}{2}}\hbar = S^*\hbar \quad (4.12)$$

In this case the magnitude is 0 or $2^{\frac{1}{2}}\hbar$, corresponding to $S = 0$ or 1.

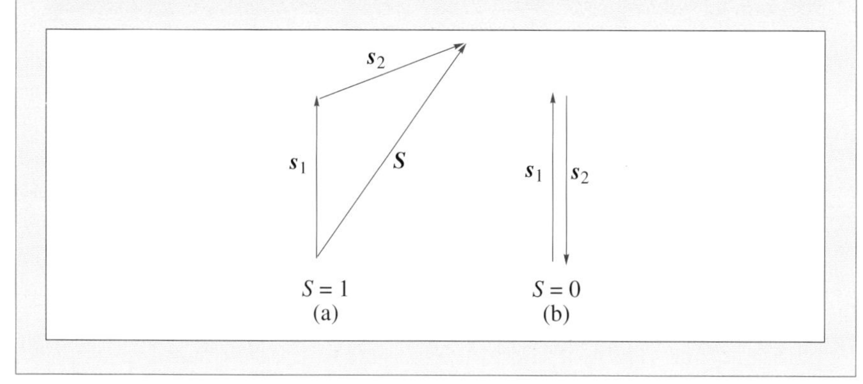


**Figure 4.5** Vector diagrams for the coupling of two electron spin angular momenta to give (a) a triplet and (b) a singlet term

For polyelectronic atoms in general, electrons in filled sub-shells can be regarded as being in pairs with antiparallel spins so that they do not contribute anything to the overall spin quantum number $S$. However, there may be more than two unpaired electrons in unfilled sub-shells. For example, the ground configuration of the phosphorus atom is:

$$\text{P} \quad [\text{Ne}]3\text{s}^2 3\text{p}^3 \quad (4.13)$$

Only the electrons in the 3p sub-shell contribute to $S$. In the case of helium we have seen that, typically, for two excited terms which differ only in their multiplicity, that of higher multiplicity lies lower in energy. For ground configurations which give rise to states of different multiplicity this is *always* the case: this is one of **Hund's rules**.

Figure 4.6(a) shows that, in the ground term of phosphorus, the three 3p electrons are distributed so that there is one in each of the $3p_x$, $3p_y$ and $3p_z$ orbitals with parallel spins. Then, $S = \frac{1}{2} + \frac{1}{2} + \frac{1}{2} = \frac{3}{2}$ and $2S + 1 = 4$, giving a quartet term. Figure 4.6(b) shows that the Pauli principle also allows two of the electrons to be in the same orbital, say $3p_x$, with antiparallel spins. For the resulting term, $S = \frac{1}{2}$, and it is a doublet term. Because Hund's rule tells us it cannot be the ground term, it must be a low-lying excited term; it is low-lying because it takes very little energy to reverse the direction of one electron spin.

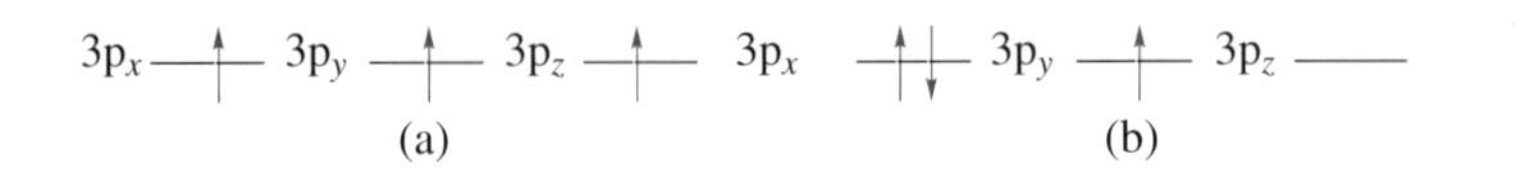


**Figure 4.6** Three electrons in 3p orbitals may give (a) a quartet or (b) a doublet term

## Worked Problem 4.4

**Q** What are the possible multiplicities which can arise from the ground configuration of the molybdenum atom?

**A** The ground configuration is:

$$\text{Mo [Kr]}4d^5 5s^1$$

The electrons in the shells with $n$ = 1, 2 and 3, and those in the 4s and 4p filled sub-shells, do not contribute to $S$. It is possible that each of the five 3d electrons can be in each of the 3d orbitals with parallel spins. Then, if the lone electron in the 5s orbital also has its spin parallel to those five:

$$S = 6 \times \tfrac{1}{2} = 3 \text{ and } 2S + 1 = 7$$

and the term is a septet term. This is the highest multiplicity possible with this configuration, and is illustrated in Figure 4.7(a).

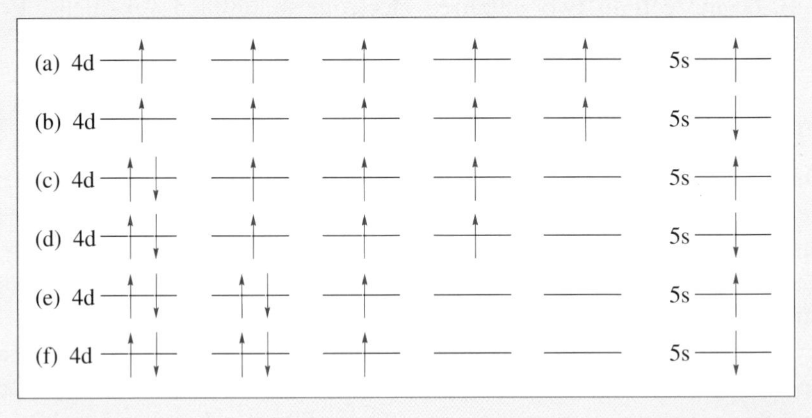


**Figure 4.7** Five electrons in a 4d and one electron in a 5s orbital may give (a) a septet, (b) a quintet, (c) a quintet, (d) a triplet, (e) a triplet or (f) a singlet term

In Figure 4.7(b), the spin of the 5s electron is reversed, so that $S$ = 2. The term is a quintet term, as is that resulting from the spin configuration in Figure 4.7(c). Those in Figure 4.7(d) and 4.7(e) have two unpaired spins, and the terms are triplets, while that in Figure 4.7(f) has no unpaired spins, and the term is a singlet.

### 4.3.3 Terms Arising From Two or More Electrons in the Same Sub-shell

Electrons in different unfilled sub-shells, for example in the configuration $2p^1 3p^1$, $3p^1 3d^1$ or $2p^1 3d^1$, are known as **non-equivalent electrons**: they have different values of either $n$ or $\ell$, or both. The coupling of the orbital angular momenta, and of the spin angular momenta, for any non-equivalent electrons can be carried out as in Sections 4.3.1 and 4.3.2. Electrons in the same unfilled sub-shell, having the same values of $n$ and $\ell$, are known as **equivalent electrons** and must be treated differently.

We consider the simplest example of two electrons in the 2p orbital, as in the ground configuration of the carbon atom, $1s^2 2s^2 2p^2$. There is no spin or orbital angular momentum contributed by the 1s or 2s electrons: it is only the 2p electrons which contribute. Both these electrons have the same set of quantum numbers $n$ (= 2), $\ell$ (= 1) and $s$ (= $\frac{1}{2}$). Consequently, the Pauli principle would be violated if they did not have different values of $m_\ell$ (= 0, ±1) or $m_s$ (= $\pm\frac{1}{2}$). In fact, there are 15 different possible combinations of $(m_\ell)_1$ and $(m_s)_1$, for one of the electrons, and of $(m_\ell)_2$ and $(m_s)_2$, for the other electron. These 15 combinations can be divided into five which constitute a $^1D$ term, nine a $^3P$ term and one a $^1S$ term.

**Box 4.3 Terms Arising from Two Equivalent p Electrons**

Even though $p^2$ represents the simplest case of equivalent electrons, the derivation of the three states which arise is inevitably lengthy. It is shown here to illustrate how such derivations are made using the various possible combinations of the quantum numbers $(m_\ell)_1$, $(m_\ell)_2$, $(m_s)_1$ and $(m_s)_2$ for electrons 1 and 2, for both of which $\ell$ = 1 and $s = \frac{1}{2}$.

| | | | | | | | | | | | | | | | |
|---|---|---|---|---|---|---|---|---|---|---|---|---|---|---|---|
| $(m_\ell)_1$ | 1 | 1 | 1 | 1 | 1 | 1 | 1 | 1 | 1 | 0 | 0 | 0 | 0 | 0 | –1 |
| $(m_\ell)_2$ | 1 | 0 | 0 | 0 | 0 | –1 | –1 | –1 | –1 | 0 | –1 | –1 | –1 | –1 | –1 |
| $(m_s)_1$ | $\frac{1}{2}$ | $\frac{1}{2}$ | $\frac{1}{2}$ | $-\frac{1}{2}$ | $-\frac{1}{2}$ | $\frac{1}{2}$ | $\frac{1}{2}$ | $-\frac{1}{2}$ | $-\frac{1}{2}$ | $\frac{1}{2}$ | $\frac{1}{2}$ | $\frac{1}{2}$ | $-\frac{1}{2}$ | $-\frac{1}{2}$ | $\frac{1}{2}$ |
| $(m_s)_2$ | $-\frac{1}{2}$ | $\frac{1}{2}$ | $-\frac{1}{2}$ | $\frac{1}{2}$ | $-\frac{1}{2}$ | $\frac{1}{2}$ | $-\frac{1}{2}$ | $\frac{1}{2}$ | $-\frac{1}{2}$ | $-\frac{1}{2}$ | $\frac{1}{2}$ | $-\frac{1}{2}$ | $\frac{1}{2}$ | $-\frac{1}{2}$ | $-\frac{1}{2}$ |
| $M_L = \Sigma_i(m_\ell)_i$ | 2 | 1 | 1 | 1 | 1 | 0 | 0 | 0 | 0 | 0 | –1 | –1 | –1 | –1 | 2 |
| $M_S = \Sigma_i(m_s)_i$ | 0 | 1 | 0 | 0 | –1 | 1 | 0 | 0 | –1 | 0 | 1 | 0 | 0 | –1 | 0 |

Rearranging the pairs of values of $M_L$ and $M_S$ in the following way:

| | | | | | | | | | | | | | | | |
|---|---|---|---|---|---|---|---|---|---|---|---|---|---|---|---|
| $M_L$ | 2 | 1 | 0 | –1 | –2 | 1 | 0 | –1 | 1 | 0 | –1 | 1 | 0 | –1 | 0 |
| $M_S$ | 0 | 0 | 0 | 0 | 0 | 1 | 1 | $\underline{1}$ | 0 | 0 | 0 | –1 | $\underline{-1}$ | –1 | 0 |
| | | | $^1D$ | | | | | | | $^3P$ | | | | | $^1S$ |

shows that they result from a $^1$D, a $^3$P and a $^1$S term.

Note that, in deriving all the possible combinations of the four $m$ quantum numbers, account is taken of the fact that the two electrons are indistinguishable. This means that, if, for example, $(m_\ell)_1$ and $(m_\ell)_2$ are the same, only one of the combinations $(m_s)_1 = \frac{1}{2}$, $(m_s)_2 = -\frac{1}{2}$ and $(m_s)_1 = -\frac{1}{2}$, $(m_s)_2 = \frac{1}{2}$ can be included.

### Worked Problem 4.5

**Q** Compare the terms which arise from two non-equivalent p electrons with those that arise from two equivalent p electrons.

**A** For two non-equivalent p electrons, for example 2p and 3p for which $\ell_1 = 1$ and $\ell_2 = 1$, $L = 1 + 1, 1 + 1 - 1, 1 - 1 = 2$, 1 and 0, from equation (4.7), giving D, P and S terms. Since $s_1 = \frac{1}{2}$ and $s_2 = \frac{1}{2}$, $S = 1$ and 0, from equation (4.10), giving triplet and singlet terms.
Therefore the total number of terms arising from two non-equivalent p electrons is:

$$^3\text{D},\ ^1\text{D},\ ^3\text{P},\ ^1\text{P},\ ^3\text{S and } ^1\text{S}$$

This compares with $^1$D, $^3$P and $^1$S (see above), which arise from two equivalent p electrons. In this case the $^3$D, $^1$P and $^3$S terms arising from non-equivalent p electrons are forbidden by the Pauli principle.

Box 4.3 shows how lengthy is the procedure necessary for deriving the terms from two equivalent p electrons. The procedure becomes even more lengthy, and the number of terms more numerous, when more electrons, and particularly d electrons, are involved. For example, a $d^2$ configuration gives rise to $^1$S, $^3$P, $^1$D, $^3$F and $^1$G terms, and a $d^3$ configuration gives rise to $^2$P, $^4$P, $^2$D (2 terms), $^2$F, $^4$F, $^2$G and $^2$H terms.

To determine which of the various terms arising from equivalent electrons, such as the $p^2$, $d^2$ and $d^3$ examples we have encountered, is the ground term we need to apply Hund's rules, which are:

1. Of the terms arising from equivalent electrons, those with the highest multiplicity lie lowest in energy.
2. Of these, the lowest is that with the highest value of $L$.

From these rules it follows that the ground term of the carbon atom ($p^2$)

is $^3P$, the ground term of the titanium atom ($d^2$) is $^3F$, and that of the vanadium atom ($d^3$) is $^4F$.

There is another useful rule regarding terms which arise from equivalent electrons. This is that a vacancy in an orbital behaves, in respect of the terms which arise, like an electron. Therefore the ground term of the oxygen atom ($p^4$) is, like that of the carbon atom ($p^2$), $^3P$. Also, the ground term of the nickel atom ($d^8$) is $^3F$, like that of the titanium atom ($d^2$).

### 4.3.4 Multiplet Structure in Atomic Spectra

We have seen how the orbital angular momenta of all the electrons in an atom are coupled to give the total angular momentum, and the spin momenta are coupled to give the total orbital and the total spin angular momenta, as illustrated by equations (4.7) and (4.10), respectively. In fact, this treatment of orbital and spin angular momenta is an approximation known as the **Russell–Saunders coupling** approximation. The approximation involved is the complete neglect of any coupling between the orbital angular momentum of an electron to its own spin angular momentum, namely the $\ell s$ coupling.

An alternative approximation is the *jj* coupling approximation. This is the opposite to the Russell–Saunders approximation in that the orbital and spin momenta of each electron are regarded as being strongly coupled, and coupling of all angular momenta, and of all the spin momenta, is neglected. The Russell–Saunders approximation is appropriate for most states of most atoms, and is the only one that we shall consider here.

In the Russell–Saunders approximation the total orbital angular momentum and the total spin angular momentum couple together to give the **total angular momentum**. The quantum number associated with this is $J$ which can take the values:

$$J = L + S, L + S - 1, \ldots |L - S| \tag{4.14}$$

**Box 4.4 Vector Representation of Total Orbital and Total Electron Spin Angular Momenta**

Vector diagrams can be drawn to illustrate the coupling of the total orbital and total spin angular momentum vectors, where the magnitudes are given by equations (4.9) and (4.12), respectively. The resultant vector is of magnitude:

$$[J(J + 1)]^{\frac{1}{2}}\hbar = J^*\hbar \tag{4.15}$$

where $J$ takes the values given by equation (4.14).

The example in Figure 4.8 shows the vector diagrams for the three states arising from a $^3$D term. For this term, $S = 1$, $L = 2$ and $J = 3$, 2 and 1. The magnitudes of the orbital and spin vectors are $6^{\frac{1}{2}}\hbar$ and $2^{\frac{1}{2}}\hbar$, respectively, and the possible magnitudes of the total angular momentum vector are $12^{\frac{1}{2}}\hbar$, $6^{\frac{1}{2}}\hbar$ and $2^{\frac{1}{2}}\hbar$.

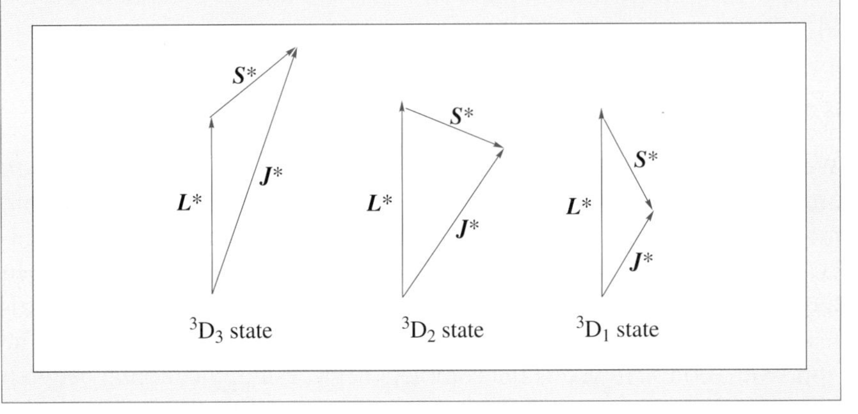


**Figure 4.8** Vector diagrams for the coupling of orbital and spin angular momenta for a $^3$D term

The number of values that $J$ can take determines the number of states that arise from a particular term. In the case of the $^3$P term, $L = 1$, $S = 1$ and therefore $J = 2$, 1 and 0. States are labelled with the value of $J$ as a post-subscript so that these three states are labelled $^3P_2$, $^3P_1$ and $^3P_0$, respectively.

## Worked Problem 4.6

**Q** What states arise from the terms (a) $^3$S, (b) $^4$P, (c) $^5$D and (d) $^2$F?

**A** (a) $L = 0$, $S = 1$; $\therefore$ $J = 1$ and the only state is $^3S_1$.

Because $L = 0$, only one state arises from all S terms, whatever the spin multiplicity.

(b) $L = 1$, $S = 3/2$; $\therefore$ $J = 5/2$, 3/2 and 1/2, giving $^4P_{5/2}$, $^4P_{3/2}$ and $^4P_{1/2}$ states.

(c) $L = 2$, $S = 2$; $\therefore$ $J = 4$, 3, 2, 1 and 0, giving $^5D_4$, $^5D_3$, $^5D_2$, $^5D_1$ and $^5D_0$ states.

(d) $L = 3$, $S = \frac{1}{2}$; $\therefore$ $J = 7/2$ and 5/2, giving $^2F_{7/2}$ and $^2F_{5/2}$ states.

In the Russell–Saunders approximation we have neglected **spin–orbit coupling**, the $\ell s$ coupling between the spin and orbital motion of the same electron. This coupling is caused by the interaction of the magnetic moments associated with the spin and orbital motions of the electron. One effect of spin–orbit coupling is to split apart the states with different values of $J$ arising from a particular term. The splitting, $\Delta E$, in terms of energy, between the $J$th and the $(J-1)$th energy levels is given by:

$$\Delta E = E_J - E_{J-1} = AJ \tag{4.16}$$

for most, but not all, **multiplet** states. The quantity $A$ is constant for a particular multiplet and, in general, may be positive or negative. The fact that $\Delta E$ is proportional to $J$ in equation (4.16) is known as the **Landé interval rule**. In general, the rule holds for atoms in which spin–orbit coupling is small. Since this coupling is proportional to $Z^4$, where $Z$ is the charge on the nucleus, the rule holds best for atoms with low $Z$, *i.e.* light atoms. The helium atom, for which the interval rule does not hold, is an important exception.

Figure 4.9 shows two examples of multiplets split by spin–orbit coupling. In Figure 4.9(a), the splitting of a $^3$P term into three states, with a positive value of $A$, has the state with the smallest value of $J$ the lowest in energy. Such a multiplet, with a positive value of $A$, is known as a **normal multiplet**. The multiplet of four states, arising from a $^4$D term and shown in Figure 4.9(b), is an **inverted multiplet** with a negative value of $A$.

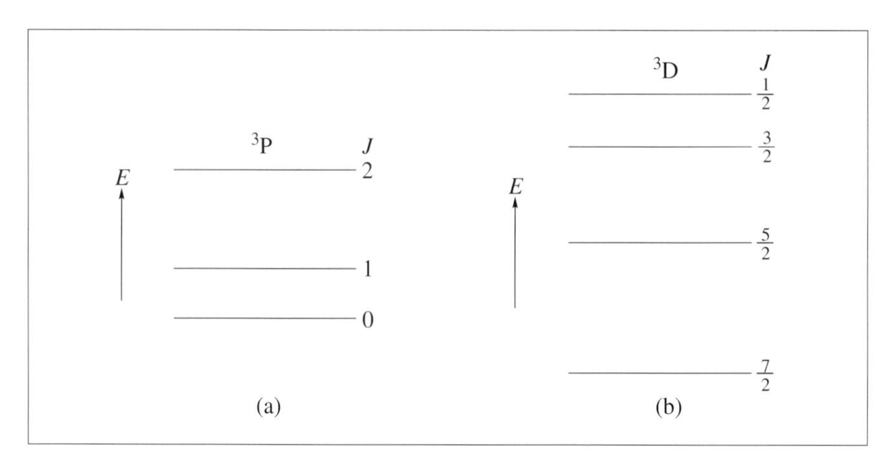


**Figure 4.9** Multiplet states arising from (a) a $^3$P and (b) a $^4$D term

There are no general rules regarding the normal or inverted nature of multiplets arising from excited terms, but there are two useful rules regarding multiplets arising from ground terms:

1. When an orbital is partially filled with equivalent electrons, and is less than half full, the multiplet that arises is normal.
2. When an orbital is partially filled with equivalent electrons, and is more than half full, the multiplet that arises is inverted.

For example, the ground $^3$P term of carbon, resulting from two equivalent p electrons, forms a normal multiplet, like that in Figure 4.9(a), and the ground state is $^3P_0$. On the other hand, the ground $^3$P term of oxygen, resulting from four equivalent p electrons, forms an inverted multiplet, and the ground state is $^3P_2$.

For atoms having a ground configuration with an exactly half-filled orbital, the ground term is always an S term, for which only one state, the ground state, can arise. For example, the ground term of the chromium atom is $^7$S but, because the quantum numbers $L = 0$ and $S = 3$, $J$ can take only the value 3, and the ground state is $^7S_3$.

**Worked Problem 4.7**

**Q** The vanadium atom has the ground configuration [Ar]3d$^3$4s$^2$. Given that a d$^3$ configuration gives rise to $^2$P, $^4$P, $^2$D (two terms), $^2$F, $^4$F, $^2$G and $^2$H terms, determine the values of $L$, $S$ and $J$ for the ground state.

**A** Hund's first rule tells us that the ground term has the highest multiplicity, and must be therefore $^4$P or $^4$F. The second rule tells us that, of these two terms, the ground term must be that with the highest value of $L$, *i.e.* $^4$F.

For the $^4$F term, $L = 3$ and $S = 3/2$. Therefore $J = 9/2$, 7/2, 5/2 and 3/2. Since the 3d orbital is less than half full, the multiplet arising from the ground term is normal, and the ground state is $^4F_{3/2}$.

## 4.4 Selection Rules in Spectra of Polyelectronic Atoms

For the relatively simple case of the helium atom, considering only the promotion of one of the 1s electrons and not taking into account the splitting of the $^3$P, $^3$D, $^3$F, ... terms into multiplets, the selection rules for transitions between terms are given in equation (4.5). For polyelectronic atoms in general, and taking into account multiplet splitting, these selection rules must be expanded as follows:

1. $\Delta n$ is unrestricted. This selection rule applies to all atomic spectra.
2. $\Delta L = 0, \pm 1$ except that $L = 0 \nleftrightarrow L = 0$, where $\nleftrightarrow$ implies that this type of transition is forbidden. This selection rule is quite general and applies to the promotion of any number of electrons.
3. $\Delta J = 0, \pm 1$, except that $J = 0 \nleftrightarrow J = 0$. This restricts transitions between components of multiplets.

4. $\Delta\ell = \pm 1$, for the promotion of a single electron. One important consequence of this rule is that transitions are forbidden between states arising from the same configuration. For example, the excited configuration $1s^22s^22p^13d^1$ of carbon gives rise to $^1F$, $^1D$, $^1P$, $^3F$, $^3D$ and $^3P$ terms. The $^3F$ term gives rise to $^3F_4$, $^3F_3$ and $^3F_2$ states, and the $^3D$ term gives rise to $^3D_3$, $^3D_2$ and $^3D_1$ states. The $^3F_3$–$^3D_2$ transition, for example, would be allowed by the $\Delta L$, $\Delta S$ and $\Delta J$ selection rules but, because both states arise from the same configuration, $\Delta\ell$ would be zero which forbids the transition.
5. $\Delta S = 0$. This selection rule must be treated with caution because it does not apply universally: it is broken down by spin–orbit coupling. We have seen in Section 4.3.4 that this coupling is proportional to $Z^4$ and, therefore, the heavier the atom, the more the selection rule tends to break down. The intensity of such **spin-forbidden** transitions depends on the extent of spin–orbit coupling. For example, no transitions between singlet and triplet states have been found in the helium atom, whereas the $6^3P_1$–$6^1S_0$ transition is one of the strongest in the emission spectrum of the mercury atom, for which $Z$ (= 80) is sufficiently high to break down completely the $\Delta S = 0$ selection rule.

It is conventional in all spectroscopy to denote a **transition** between an upper state B and a lower state A by B–A. If it is helpful to indicate that the transition is in absorption or emission, the symbolism B ← A or B → A, respectively, may be used.

Like all selection rules, these apply to either absorption or emission of radiation, although, usually, electronic spectra of atoms are observed in emission.

The sodium atom, like all alkali metal atoms, has a relatively simple ground configuration, with one electron, the valence electron, in an outer s orbital:

$$\text{Na} \quad 1s^22s^22p^63s^1 \tag{4.17}$$

and the ground state is $3^2S_{1/2}$. Promotion of the 3s electron to the lowest p orbital, the 3p orbital, results in two states, $3^2P_{3/2}$ and $3^2P_{1/2}$. In this case the $^2P$ multiplet is normal, having the state with the higher value of $J$ higher in energy. Figure 4.10(a) shows these two states, and the two transitions between them and the ground state which are allowed by the selection rules. Comparison of this figure with Figure 1.3 shows that the doublet structure of the sodium D lines, which occur with very high intensity in the emission spectrum of sodium and are responsible for the predominantly yellow colour of a sodium discharge lamp, is due to spin–orbit coupling which splits the $6^2P_{3/2}$ and $6^2P_{1/2}$ states. All $^2P$–$^2S$ transitions show similar doublet structure.

Figure 4.10(b) shows how a $^2D$–$^2P$ transition, in this case $6^2D$–$6^2P$, is split into a triplet by spin–orbit coupling. The general rule that spin–orbit splitting decreases as $L$ increases is reflected in the smaller splitting shown in the $^2D$ than in the $^2P$ multiplet.

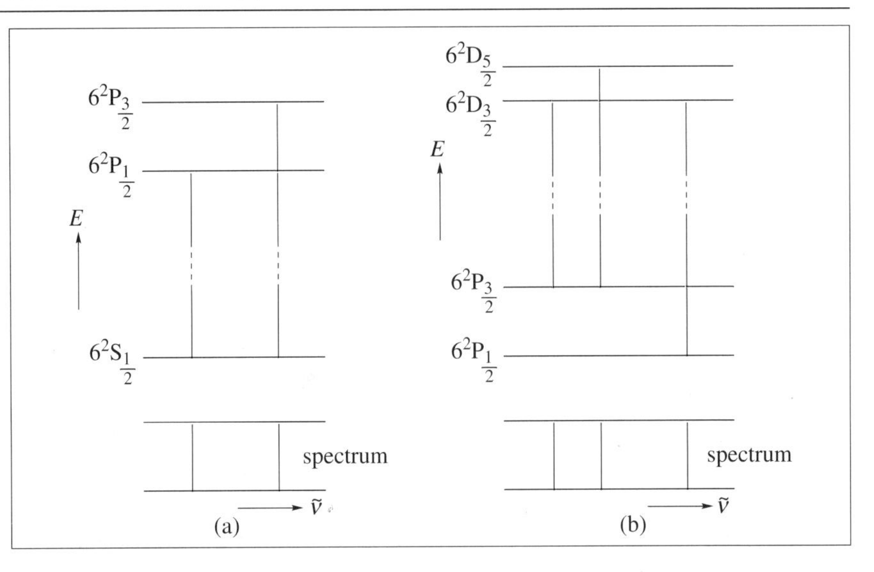


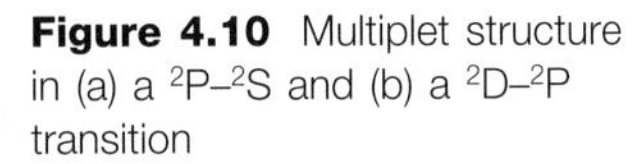

**Figure 4.10** Multiplet structure in (a) a $^2P$–$^2S$ and (b) a $^2D$–$^2P$ transition

## Summary of Key Points

**1.** *Repulsion between electronic charges*
Even in the helium atom, the effect of electron–electron repulsion is to split the degeneracy of, for example, the 2s and 2p orbitals, and of the 3s, 3p and 3d orbitals, in the hydrogen atom.

**2.** *The aufbau and Pauli exclusion principles*
For the ground configuration of any atom, electrons are fed into the orbitals in order of increasing energy, making sure that no two electrons have the same set of quantum numbers $n$, $\ell$, $m_\ell$ and $m_s$.

**3.** *Orbital picture of the Periodic Table*
How the aufbau and Pauli principles lead to consistent pictures of, for example, the noble gas, alkali metal, alkaline earth metal and transition metal atoms.

**4.** *The spectrum of the helium atom*
Manifolds of singlet and triplet states. Selection rules.

**5.** *Coupling of orbital and spin angular momenta in polyelectronic atoms*
$\ell\ell$, $ss$ and $LS$ coupling in the Russell–Saunders approximation.

**6.** *Coupling of orbital and spin angular momenta of equivalent electrons*
Necessity of deriving all possible combinations of the quantum

numbers $m_\ell$ and $m_s$, taking into account the indistinguishability of the electrons.

7. *Multiplet structure in atomic spectra*
Doublet, triplet, ... terms with $L > 0$ split into multiplet states. Landé interval rule. Selection rules.

## Problems

**4.1.** What are the ground states ($L$, $S$ and $J$ values) of the (a) nitrogen, (b) phosphorus and (c) manganese atoms?

**4.2.** Derive the ground state of the nickel atom and compare it with that of titanium.

# 5

# Electronic States of Diatomic and Polyatomic Molecules

**Aims**

In this chapter you will be introduced to:

- The method of linear combination of atomic orbitals (LCAO) for obtaining molecular orbitals from atomic orbitals
- Its application to homonuclear diatomic molecules
- Its application to heteronuclear diatomic molecules
- Its application to polyatomic molecules
- Derivation of electronic states from ground molecular orbital configurations
- Derivation of electronic states from excited molecular orbital configurations

## 5.1 Homonuclear Diatomic Molecules

### 5.1.1 Molecular Orbitals and Ground Electron Configurations

In Chapter 4 we moved from the comparative simplicity of the spectrum of the hydrogen atom, and its interpretation, to helium and other polyelectronic atoms, in which the presence of more than one electron introduces electron–electron repulsions. The effects of the presence of more than one electron and of these repulsions are far-reaching. In experimental terms, the simple elegance of the hydrogen atom spectrum is lost and, in theoretical terms, the calculation of the energy levels and wave functions becomes increasingly approximate as the number of electrons increases.

In diatomic molecules, a further electrostatic interaction is introduced, that of repulsion between the positively charged nuclei. For all except the simplest molecules, particularly $H_2^+$ and $H_2$, which have only one or two electrons, the calculation of energy levels and wave functions becomes still more difficult. Here, we shall be concerned only with pictorial representations of wave functions and the relative ordering of energy levels, and shall treat them only semi-quantitatively.

There are two approximate theoretical approaches, using **molecular orbital theory** or **valence bond theory**. However, it is molecular orbital theory which is the most generally useful, and it is only this that we shall use here.

In a diatomic molecule an electron is said to be in a **molecular orbital** (MO). These MOs are constructed from atomic orbitals (AOs) by assuming that, in a homonuclear diatomic molecule, $A_2$, the MO resembles an AO of atom A when the electron is in the vicinity of either of the nuclei. This is taken account of in the construction of MO wave functions, at the lowest level of approximation, by the method of **linear combination of atomic orbitals** (LCAO).

In general, for any diatomic molecule, the LCAO method gives MO wave functions, $\psi$, of the form:

$$\psi = c_1\chi_1 + c_2\chi_2 \tag{5.1}$$

where $\chi_1$ and $\chi_2$ are AO wave functions for atoms 1 and 2, and $c_1$ and $c_2$ are constants reflecting the proportions of $\chi_1$ and $\chi_2$ which constitute the MO. In the case of a homonuclear diatomic molecule, $\chi_1 = \chi_2$ and $c_1 = \pm c_2$, and we have:

$$\psi = N(\chi_1 \pm \chi_2) \tag{5.2}$$

where $N$ replaces $c$ and is a **normalization constant** whose value will not concern us here.

## Box 5.1 Normalization of a Wave Function

The necessity of a normalization constant in all wave functions $\psi$ is to ensure that the normalization condition:

$$\int \psi^2 \, d\tau = 1 \tag{5.3}$$

is satisfied, where $d\tau$ indicates that the integration is over all space. The quantity $\psi^2$ is the **probability** of finding the electron at a particular point in space, and the normalization condition in equation (5.3) simply makes certain that the probability of finding the electron anywhere in space is one.

Figure 5.1 shows the simplest case of MO wave functions arising from 1s AOs on each atom. In Figure 5.1(a) the MO wave function illustrated is $N(\chi_{1s} + \chi_{1s})$. Since the electron charge density at any point in space is proportional to $\psi^2$, it is high in the **overlap region** of the AOs and therefore the MO is a **bonding orbital**. The MO wave function shown in Figure 5.1(b) is $N(\chi_{1s} - \chi_{1s})$, in which the wave functions cancel in the region midway between the nuclei. The region in which $\psi = 0$, and in which there is zero electron density, is called a **nodal plane** and the MO is **antibonding**. The linear combination in Figure 5.1(a) is an **in-phase** combination and that in Figure 5.1(b) an **out-of-phase** combination.

There are two important rules which must be obeyed in the LCAO method:

1. The AOs being combined must have the same symmetry with respect to the internuclear axis. For example, a linear combination of a 2s and a $2p_x$ AO, illustrated in Figure 5.2(a), is not allowed. This figure shows that, if the nuclei were moved closer together in an attempt to form a bond, any overlap of the orbitals in the upper half of the figure would be exactly cancelled by that in the lower half.

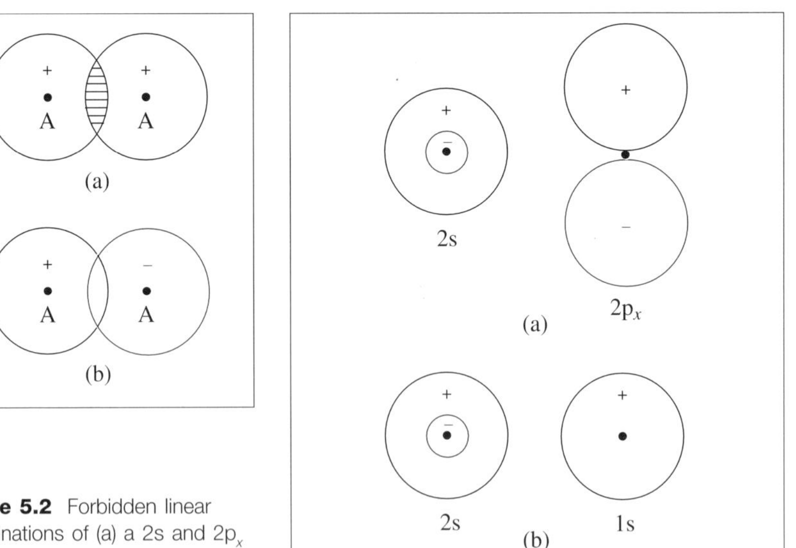


**Figure 5.1** (a) In-phase and (b) out-of-phase combinations of 1s AO wave functions to form a bonding and antibonding MO, respectively

**Figure 5.2** Forbidden linear combinations of (a) a 2s and $2p_x$ and (b) a 1s and a 2s AO

When we say that AOs have the same **symmetry** with respect to the internuclear axis, we mean that they have cylindrical symmetry about that axis, *i.e.* the wave function is unchanged after rotation by any angle about the axis. For example, all *n*s orbitals, and all $np_z$ orbitals directed along the internuclear *z*-axis, are symmetric to this rotation, whereas all $np_x$ and $np_y$ orbitals are not.

2. The AOs being combined must have the same energy. For example, although the 1s and 2s AOs in Figure 5.2(b) have the same symmetry with respect to the internuclear axis, they have very different energies and do not form MOs.

Solution of the Schrödinger equation (equation 3.19) with LCAO wave functions of the type given in equation (5.2) gives the energies, $E_\pm$, associated with the two MOs as:

$$E_{\pm} = (E_A \pm \beta)/(1 \pm S) \qquad (5.4)$$

where $E_A$ is the energy associated with the AO on atom A. The relative energies of the two resulting MOs are shown in Figure 5.3. The quantity $\beta$ is the **resonance energy** and, as the figure shows, is a negative quantity. The two MOs are separated in energy by approximately $2\beta$. The magnitude of the resonance energy is an indicator of the difference between the AOs and the MOs which are formed from them. For example, if we had tried to form MOs from a 1s AO on one atom and a 2s AO on the other, we would have found that, although the AOs have the same symmetry with respect to the internuclear axis, the resonance energy $\beta = 0$ because the AOs are of very different energies.

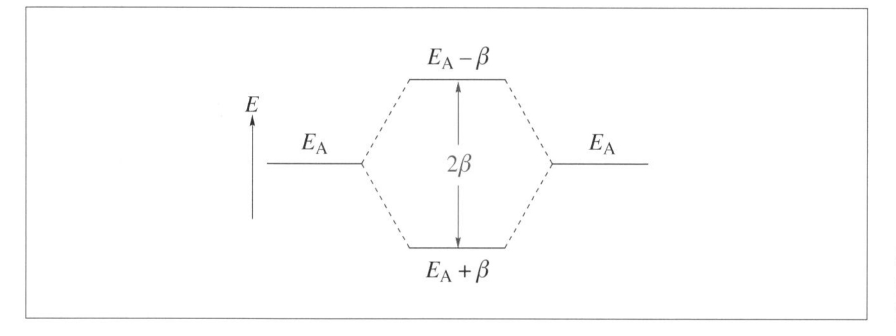


**Figure 5.3** Energy levels for two MOs resulting from LCAO treatment of two AOs

The quantity $S$ in equation (5.4) is the **overlap integral**. As the name implies, it is a measure of the extent to which the two AOs overlap. For example, the hatched area in Figure 5.1(a) shows a region of overlap between two 1s AOs. For total overlap, $S = 1$, but typically $S \approx 0.2$. Although this value is far from negligible, it can be neglected in a very approximate calculation of MO energies. Equation (5.4) then becomes:

$$E_{\pm} \approx E_A \pm \beta \qquad (5.5)$$

and the MO energies are symmetrically disposed abut the AO energy $E_A$, as shown in Figure 5.3.

Figure 5.4 shows pictorial representations of MOs formed from 1s, 2s and 2p AOs. In every case in which we make a linear combination of AOs, two MOs result. As with a linear combination of 1s AOs, shown in Figure 5.1, there is always an in-phase and an out-of-phase combination, resulting, respectively, in a bonding and an antibonding MO. In the figure the corresponding wave functions are labelled $\psi^+$ and $\psi^-$. The antibonding MO is always higher in energy than the bonding MO. This is a consequence of there being an additional nodal plane in the antibonding orbital. The nodal plane is halfway between the nuclei and

"Normal", in this context, means that the plane is at right angles, *i.e.* at 90°, to the figure.

normal to the figure in all cases. There is a general rule that the introduction of a nodal plane increases the energy of the MO.

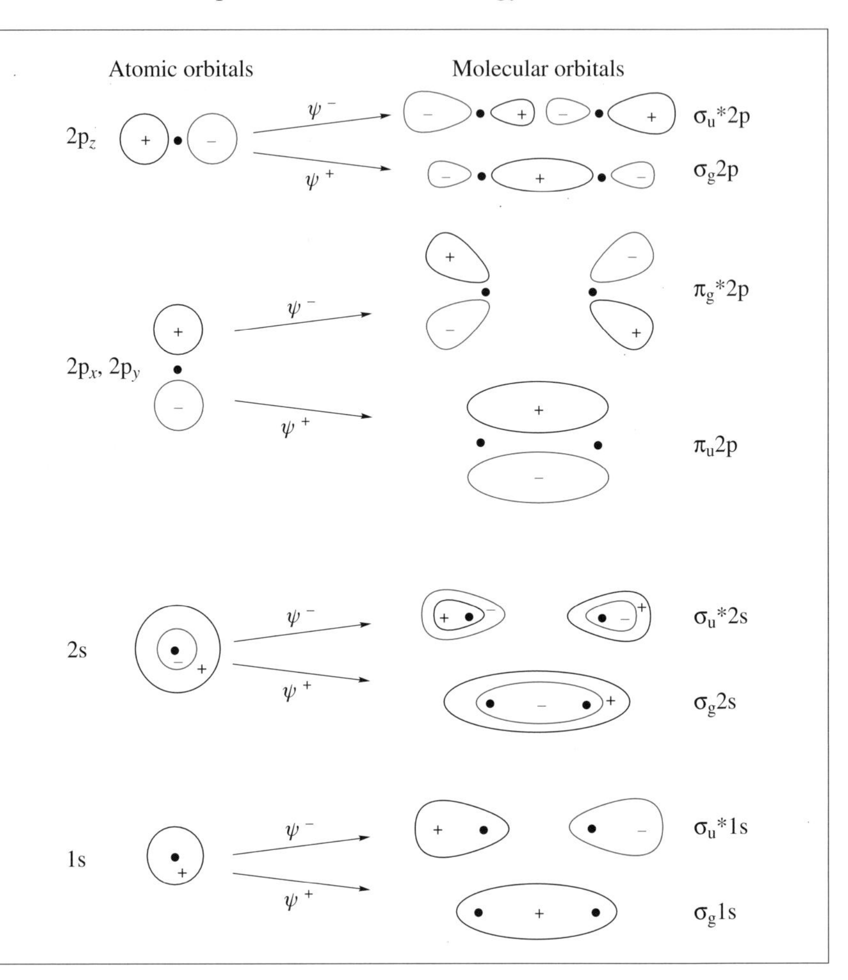


**Figure 5.4** MOs formed from 1s, 2s and 2p AOs for a homonuclear diatomic molecule

There are four components of the labels attached to the MOs:

1. The AOs from which the MOs are formed, in these cases 1s, 2s and 2p.
2. The σ or π designation. This can be regarded as indicating the symmetry of the MO with respect to rotation around the internuclear axis. The σ MOs are cylindrically symmetric, *i.e.* the MOs are unchanged by any rotation about the axis, whereas the π MOs are not: they have one nodal plane through the *z*-axis. The σ and π MOs are also distinguished by their degeneracy. The σ MOs are non-degenerate whereas the π MOs are doubly degenerate. Double degeneracy implies that there are two different MOs which have different wave functions but the same energy. For example, as well as the $\pi_u2p$

MO shown in Figure 5.4, there is an otherwise identical one which is in a plane normal to the figure. If the one in the figure is formed from $2p_x$ AOs, the one normal to the figure is formed from $2p_y$ AOs.

3. Those MOs marked with an asterisk (*) are antibonding while those which are unmarked are bonding.
4. The subscripts g and u, standing for "gerade" (even) and "ungerade" (odd), indicate that an MO is either symmetric or antisymmetric, respectively, to inversion through a point at the centre of the molecule. The operation of inversion involves going from any point in the molecule to another point an equal distance on the opposite side of the centre point. If, as a result of inversion, the MO changes sign, it is a "u" MO; if it does not, it is a "g" MO.

The use of *both* * to indicate an antibonding MO, and g or u to indicate symmetry or antisymmetry to inversion, is unnecessary. The former is useful only when we need reminding that the MO is antibonding, and will not be used subsequently for homonuclear diatomic molecules.

Figure 5.5 shows an energy level diagram for the MOs formed from 1s, 2s and 2p AOs. The diagram shows that an antibonding orbital is always higher in energy than the corresponding bonding orbital, and has been drawn showing the $\sigma_g 2p$ orbital higher in energy than the $\pi_u 2p$ orbital. Although the relative energies are only semi-quantitative, the diagram has been drawn so that the two $\sigma 2s$ and the two $\sigma 2p$ MO energy levels are not quite symmetrically disposed about the 2s and 2p AO energy levels, respectively. The reason for the asymmetry is that the $\sigma_g 2s$ and $\sigma_g 2p$ MOs would, otherwise, be fairly close together. Because they have the same symmetry, they push each other apart, thereby losing the symmetrical disposition of the bonding and antibonding energy levels of equation (5.5). For diatomic molecules consisting of atoms from the first and second periods of the Periodic Table (H to Ne), this is the case for almost all of them. Important exceptions are $O_2$ and $F_2$. In these two

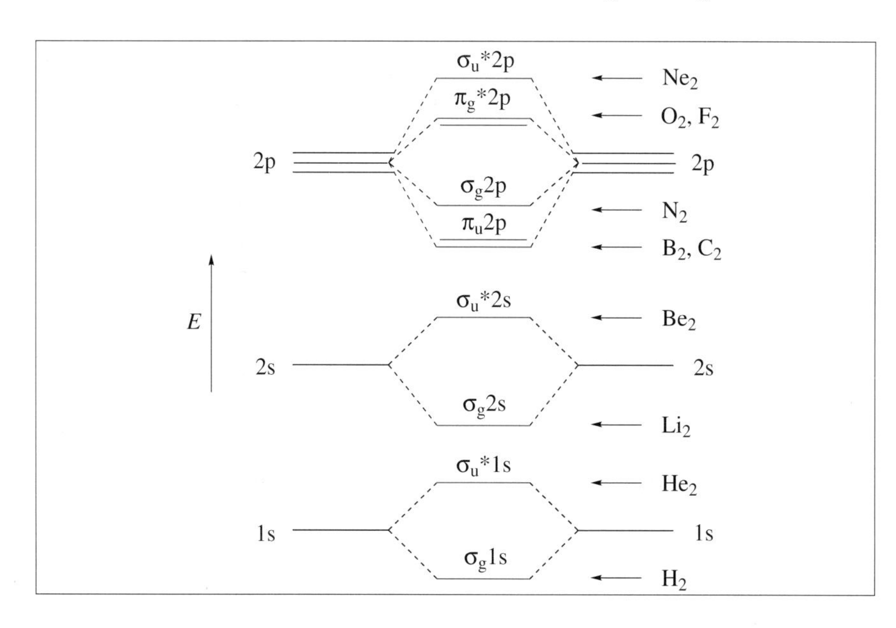


**Figure 5.5** Energy level diagram for MOs formed from 1s, 2s and 2p AOs for a homonuclear diatomic molecule

molecules the pushing apart of the $\sigma_g2s$ and $\sigma_g2p$ MOs is much less pronounced, and the order of the $\sigma_g2p$ and $\pi_u2p$ MOs is reversed compared to that in Figure 5.5.

To obtain the electronic ground configuration of a particular molecule, all the available electrons are fed into the orbitals, in order of increasing energy. This procedure employs the aufbau principle, as in atoms (see Section 4.1). Each $\sigma$ orbital is non-degenerate and may accommodate two electrons. They have antiparallel spins (see Section 4.3.2) and therefore do not make any contribution to the overall electron spin quantum number *S*. The $\pi$ orbitals are doubly degenerate and can accommodate four electrons; their spins comprise two antiparallel pairs, and also do not contribute to *S*.

For the ground configuration of $H_2$, both electrons are in the $\sigma 1s$ orbital, giving the configuration $(\sigma_g1s)^2$. Since both electrons are in a bonding orbital, $H_2$ is a stable molecule in its ground state. We know that the bond is a single bond. Alternatively, we say that the **bond order** is one. In general, the definition of bond order is:

$$\text{Bond order} = (\text{net number of bonding electrons}) \div 2 \qquad (5.6)$$

The cation $H_2^+$ has only one electron in the $\sigma_g1s$ orbital. It is also a stable molecule in its ground state but has a weak bond; the bond order is $\frac{1}{2}$.

We would not expect $He_2$ to have a stable ground configuration since He is a noble gas. The molecule has four electrons, and the ground configuration is $(\sigma_g1s)^2(\sigma_u1s)^2$. It is a general rule that the bonding characteristics of an electron in a bonding orbital are cancelled by the antibonding characteristics of an electron in an antibonding orbital. The net result therefore is that, overall, the four electrons are non-bonding and the molecule is not stable. It is interesting to note, however, that, if an electron is promoted from the $\sigma_u1s$ to the $\sigma_g2s$ orbital, the resulting excited configuration $(\sigma_g1s)^2(\sigma_u1s)^1(\sigma_g2s)^1$ gives a stable molecule with a bond order of one. This is confirmed by the observed electronic spectra of $He_2$. Similarly, we can see that $H_2$ is unstable in the excited configuration $(\sigma_g1s)^1(\sigma_u1s)^1$.

In Figure 5.5 are indicated the highest energy orbitals (sometimes called the **outer orbitals**) which are fully or partially occupied in the ground configurations of all homonuclear diatomic molecules formed from first-row atoms. For example, in $N_2$ there are 14 electrons, and the ground configuration is:

$$N_2 \quad (\sigma_g1s)^2(\sigma_u1s)^2(\sigma_g2s)^2(\sigma_u2s)^2(\pi_u2p)^4(\sigma_g2p)^2 \qquad (5.7)$$

There are six (net) electrons in bonding orbitals, giving a bond order of three, consistent with the triple bond with which we are familiar.

In $O_2$, there are 16 electrons, and the ground configuration is:

$$O_2 \quad (\sigma_g 1s)^2(\sigma_u 1s)^2(\sigma_g 2s)^2(\sigma_u 2s)^2(\sigma_g 2p)^2(\pi_u 2p)^4(\pi_g 2p)^2 \qquad (5.8)$$

The net number of bonding electrons is four, giving a bond order of two, consistent with a double bond. However, the two electrons in the outer, doubly degenerate, $\pi_g 2p$ orbital may go either into the same orbital with antiparallel spins, giving $S = 0$, or into different orbitals with parallel spins, giving $S = 1$. These give, respectively, a singlet or a triplet term, as in the case of atoms (see Section 4.2 on the helium atom, for example). As for atoms, one of Hund's rules applies. This tells us that, of the terms that arise from a ground electron configuration, that with the highest multiplicity lies lowest in energy. Therefore the ground term of $O_2$ is a triplet. One result of this is that $O_2$ is **paramagnetic**. This magnetic character can be demonstrated by observing a stream of liquid oxygen flowing between the poles of a large magnet. The stream tends to be drawn into the magnetic field, whereas a **diamagnetic** material would deviate away from the field.

## Worked Problem 5.1

**Q** Write down the ground configurations of the following molecules: $Li_2^+$, $B_2$, $C_2$, $C_2^+$ and $F_2$. In each case, what is the multiplicity (= $2S + 1$) of the ground term, and what is the bond order?

**A** The ground configurations are as follows:

| | $\sigma_g 1s$ | $\sigma_u 1s$ | $\sigma_g 2s$ | $\sigma_u 2s$ | $\pi_u 2p$ | $2S + 1$ | Bond order |
|---|---|---|---|---|---|---|---|
| $Li_2^+$ | 2 | 2 | 1 | | | 2 | $\frac{1}{2}$ |
| $B_2$ | 2 | 2 | 2 | 2 | 2 | 3 | 1 |
| $C_2$ | 2 | 2 | 2 | 2 | 4 | 1 | 2 |
| $C_2^+$ | 2 | 2 | 2 | 2 | 3 | 2 | $1\frac{1}{2}$ |

| | $\sigma_g 1s$ | $\sigma_u 1s$ | $\sigma_g 2s$ | $\sigma_u 2s$ | $\sigma_g 2p$ | $\pi_u 2p$ | $\pi_g 2p$ | $2S + 1$ | Bond order |
|---|---|---|---|---|---|---|---|---|---|
| $F_2$ | 2 | 2 | 2 | 2 | 2 | 4 | 4 | 1 | 1 |

In $Li_2^+$ there is one unpaired electron spin, so that $S = \frac{1}{2}$. $B_2$ is rather like $O_2$ in that Hund's rule dictates that, in the ground configuration, the two electrons in the $\pi_u 2p$ MO have parallel spins, resulting in $S = 1$. In $C_2^+$, two of the electrons in the $\pi_u 2p$ MO have antiparallel spins, leaving one unpaired; therefore $S = \frac{1}{2}$. In $C_2$ and $F_2$ there are no unpaired electron spins, so $S = 0$. Note the different order of MO energies in $F_2$.

### 5.1.2 Ground and Excited Electronic States

In Section 4.3.4 we have seen that the Russell–Saunders coupling approximation is appropriate to the derivation of most states of most atoms. In this approximation the orbital angular momenta of all the electrons are coupled to give the total angular momentum, and all the spin momenta are coupled to give the total spin angular momentum. For each electron, coupling between its own orbital and spin angular momenta is neglected. In molecules, also, the Russell–Saunders coupling approximation applies in most cases, and it is only that approximation that we shall consider here.

Associated with each electron in the molecule is an orbital angular momentum quantum number $\ell$. There is a correspondence, as follows, between the value of $\ell$ and the label used for the MO:

$$\begin{array}{lcccc} \ell & 0 & 1 & 2 & \ldots \\ & \sigma & \pi & \delta & \ldots \end{array} \tag{5.9}$$

The symbols σ, π, δ ... (sigma, pi, delta, ...) are the Greek equivalents of s, p, d ... used for atoms. As in atoms, the orbital angular momenta are strongly coupled to give the total orbital angular momentum, and the electron spin angular momenta are coupled to give the total spin angular momentum with the associated quantum number $S$.

The presence of the two positively charged nuclei introduces an electrostatic field. If they do not have a high nuclear charge, coupling between the total orbital and total spin angular momenta, spin–orbit coupling, can be neglected. This is the **Hund's case (a) coupling approximation** in which the total orbital and total spin momenta are coupled, independently, to the internuclear axis. The coupling of the orbital angular momentum to the axis is very strong, and one important effect of this is that $L$ is not a "good" quantum number, *i.e.* it does not have integer values. Because the spin angular momentum is not so strongly coupled to the axis, $S$ remains a good quantum number.

The introduction of a second nucleus, compared to the situation in an atom, introduces two new quantum numbers, $\Lambda$ and $\Sigma$. The component of the orbital angular momentum along the internuclear axis may take only the values $\Lambda\hbar$, where $\Lambda$ can take the values:

$$\begin{array}{lcccccc} \Lambda = & 0 & 1 & 2 & 3 & 4 & \ldots \\ & \Sigma & \Pi & \Delta & \Phi & \Gamma & \ldots \end{array} \tag{5.10}$$

The upper case symbols Σ, Π, Δ, .... (sigma, pi, delta, ...) are the Greek equivalents of the symbols S, P, D, ... used for states of atoms.

The component of the spin angular momentum may take only the values $\Sigma\hbar$, where $\Sigma$ can take the values:

$$\Sigma = S, S - 1, \ldots, -S \quad (5.11)$$

### Box 5.2 Vector Representation of Orbital and Electron Spin Angular Momenta in a Diatomic Molecule

In a way similar to the pictorial representation of the various angular momenta of atoms in the form of vector diagrams, shown in

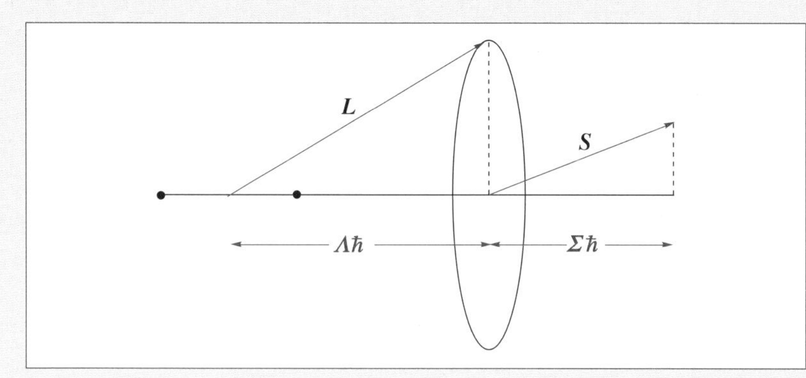


**Figure 5.6** Vector diagram for Hund's case (a) coupling of total orbital and total spin angular momenta to the internuclear axis of a diatomic molecule

Figures 4.4, 4.5 and 4.8, we can indicate the significance of the quantum numbers $\Lambda$ and $\Sigma$ in the form of vector diagrams. Figure 5.6 shows the components of orbital and spin angular momenta along the internuclear axis. The strong coupling of the angular momentum to the internuclear axis causes the corresponding vector to rotate rapidly around the axis, as indicated; this process is referred to as a **precession**. Rapid precession results in the length of the vector being ill-defined and the loss of the quantum number $L$.

Precession of the spin angular momentum is not indicated as it is relatively slow, and $S$ remains a good quantum number.

The symbolism used for electronic states is of the form ${}^{2S+1}\Lambda_{g/u}$, for example ${}^1\Sigma_g$, ${}^3\Pi_u$ and ${}^2\Delta_g$.

To obtain the electronic states that arise from a particular configuration is not always straightforward, and we shall deal with only a few cases.

Simple examples are the ground configurations of $H_2^+$ and $H_2$. $H_2^+$ has just one electron in the $\sigma_g 1s$ MO. In a $\sigma$ MO there is no orbital angular momentum and therefore $\Lambda = 0$. Since the MO is a g orbital the state is a $\Sigma_g$ state. There is only one electron; therefore $S = \frac{1}{2}$ and the state is a doublet state, ${}^2\Sigma_g$. To complete the symbolism, we need to know whether the MO wave function is symmetric or antisymmetric (changes sign) on reflection through any plane containing the internuclear axis. If it is symmetric, we add a post-superscript "+"; if it is antisymmetric, we

add "–". Since the $\sigma_g1s$ MO is cylindrically symmetrical, it is a "+" orbital and the ground state of $H_2^+$ is $^2\Sigma_g^+$. Similarly, the ground state of $H_2$, in which the two electron spins are antiparallel, is $^1\Sigma_g^+$.

As for atoms, when we derive states from configurations, filled orbitals can be ignored; they do not make any contribution to either the total spin or total orbital angular momentum. Therefore the ground state of $Li_2^+$ (see Worked Problem 5.1 for this, and other, ground configurations) is, like that of $H_2^+$, $^2\Sigma_g^+$. $F_2$ has all MOs filled and therefore has a $^1\Sigma_g^+$ ground state.

The case of the ground configuration of $B_2$ is not so straightforward. The only contributors to the overall orbital and spin angular momenta are the two electrons in the $\pi_u2p$ MO. The case of two electrons in the same $\pi$ orbital is analogous to the case in an atom of two electrons in the same p orbital, the complexity of which is illustrated in Box 4.3 in Chapter 4. Derivation of the states arising from a $(\pi_u)^2$ or $(\pi_g)^2$ configuration requires an advanced use of symmetry, which is beyond the scope of this book. We shall require only the results, which are as follows:

| Configuration | States arising | |
|---|---|---|
| $(\pi_g)^2$ | $^3\Sigma_g^- + {}^1\Sigma_g^+ + {}^1\Delta_g$ | |
| $(\pi_u)^2$ | $^3\Sigma_g^- + {}^1\Sigma_g^+ + {}^1\Delta_g$ | (5.12) |

Two simple symmetry rules which have been used in obtaining these results are:

$$g \times g = g \quad \text{and} \quad u \times u = g \tag{5.13}$$

These rules become clear when we remember that g and u indicate symmetry and antisymmetry to inversion through the centre of the molecule.

It is a useful shorthand notation to use, for example, $[Be_2]$ to indicate filled MOs, as in the ground configuration of $B_2$.

We can see now that the states that arise from the $[Be_2](\pi_u2p)^2$ ground configuration of $B_2$ are $^3\Sigma_g^-$, $^1\Sigma_g^+$ and $^1\Delta_g$. Application of Hund's rule, which applies to molecules as well as atoms (see Section 4.3.2), tells us that the state of highest multiplicity lies lowest in energy. Therefore the ground state is $^3\Sigma_g^-$, and the other two states are low-lying excited states.

## Worked Problem 5.2

**Q** What electronic states arise from the ground configurations of (a) $C_2^+$, (b) $B_2^-$, (c) $F_2^+$ and (d) $O_2$?

**A** The ground configurations are given either in the Worked Problem 5.1 or in equation (5.8).

(a) $C_2^+$ has three electrons in the $\pi_u2p$ MO, the only unfilled MO. In order to derive the ground state, we can treat the single vacan-

cy in this MO as a single electron, as we did for atoms. The result is a $^2\Pi_u$ ground state.
(b) $B_2^-$ also has three electrons in the $\pi_u 2p$ MO. As for $C_2^+$, the ground state is $^2\Pi_u$.
(c) $F_2^+$ has three electrons in the $\pi_g 2p$ MO. The vacancy in this MO can be treated like a single electron. The ground state is therefore $^2\Pi_g$.
(d) Like $B_2$, $O_2$ has two electrons in a $\pi$-type MO, in this case $\pi_g 2p$, and can be treated in a similar way. The three states that arise are $^3\Sigma_g^-$, $^1\Sigma_g^+$ and $^1\Delta_g$. Hund's rule tells us that the state of highest multiplicity lies lowest in energy. Therefore the ground state is $^3\Sigma_g^-$.

If one of the electrons in the $\pi_u 2p$ MO of $B_2$ is promoted to the $\sigma_g 2p$ MO to give the excited configuration:

$$B_2 \qquad [Be_2](\pi_u 2p)^1(\sigma_g 2p)^1 \tag{5.14}$$

there are two states that arise, a singlet and a triplet state, depending on whether the two unpaired electron spins are antiparallel or parallel, respectively. Using the rule in equation (5.13), we can see that both states must be u states. Since only the $\pi_u 2p$ electron contributes to the overall orbital angular momentum (see equation 5.9), both states must be $\Pi$ states. Therefore the two states are $^1\Pi_u$ and $^3\Pi_u$. Hund's rule applies only to states arising from a ground configuration; therefore we cannot say which state will have the lower energy.

The excited configuration:

$$B_2 \qquad [Be_2](\pi_u 2p)^1(\pi_g 2p)^1 \tag{5.15}$$

gives rise to several states. Using a further symmetry rule that:

$$g \times u = u \tag{5.16}$$

tells us that they are all u states. The two electron spins may be antiparallel or parallel, giving singlet and triplet states. Similar to the case of two electrons in the same $\pi$-type MO (see equation 5.12), two electrons in different $\pi$-type MOs give rise to $\Sigma^-$, $\Sigma^+$ and $\Delta$ states. Therefore all the states arising from the configuration in equation (5.15) are $^1\Sigma_u^-$, $^1\Sigma_u^+$, $^1\Delta_u$, $^3\Sigma_u^-$, $^3\Sigma_u^+$ and $^3\Delta_u$.

Although the states arising from various electron configurations can be derived in this way, it does not tell us anything about their relative energies. Much more sophisticated, state-of-the-art MO calculations are needed to do that.

**Worked Problem 5.3**

**Q** What states arise from (a) the $[Be_2](\pi_u 2p)^3(\sigma_g 2p)^1$ and (b) the $[Be_2](\pi_u 2p)^3(\pi_g 2p)^1$ electron configurations of $C_2$?

**A** (a) The vacancy in the $\pi_u 2p$ MO can be treated like a single electron. The spin of this and the $\sigma_g 2p$ electron may be antiparallel or parallel, giving a singlet and a triplet state. Since $u \times g = u$, they must both be u states. The only orbital angular momentum is that contributed by the $\pi_u 2p$ vacancy, and both states are $\Pi$ states. The two states are therefore $^1\Pi_u$ and $^3\Pi_u$.
(b) Again, the $\pi_u 2p$ vacancy can be treated like a single electron. The states arising are the same as those from $[Be_2](\pi_u 2p)^1(\pi_g 2p)^1$. This is the same as the $B_2$ configuration in equation (5.15), and the states arising are the same, $^{1,3}\Sigma_u^-$, $^{1,3}\Sigma_u^+$ and $^{1,3}\Delta_u$.

## 5.2 Heteronuclear Diatomic Molecules

### 5.2.1 Molecular Orbitals and Ground Electron Configurations

Heteronuclear diatomic molecules comprise a wide range of types. The simplest, at this level of approximation, are those such as CN and NO which have sufficiently similar nuclear charges for the molecules to be treated in a way rather similar to that for homonuclear diatomics. At the other extreme are molecules such as HCl, in which the nuclei differ substantially.

For heteronuclear diatomic molecules containing atoms from the second period of the Periodic Table, and formed from atoms with similar nuclear charges, it is useful to assume a set of MOs similar to those for homonuclear diatomics in Figure 5.5. There are three main differences:

1. The g or u symmetry with respect to inversion through the centre of the molecule is lost.
2. The energies of the two 1s AOs, the two 2s AOs, *etc.*, are now slightly different; the greater the difference in the nuclear charges of the two atoms, the greater is this energy difference.
3. Because the g/u symmetry is lost, it is now essential to include the (*) notation to indicate an antibonding orbital in order to distinguish it from the corresponding bonding orbital.

With the order of MO energies in Figure 5.5, the corresponding MOs for different nuclei are, in order of increasing energy:

$$\sigma 1s < \sigma^* 1s < \sigma 2s < \sigma^* 2s < \pi 2p < \sigma 2p < \pi^* 2p < \sigma^* 2p \quad (5.17)$$

The formation of these MOs from AOs follows the rules which apply to homonuclear diatomic molecules. These rules are that the AOs, used in the LCAO method to form MOs, must (a) have the same symmetry with respect to rotation about the internuclear axis and (b) have comparable energies.

The CO molecule has 14 electrons; it is **isoelectronic** with, *i.e.* has the same number of electrons as, $N_2$. In the ground configuration the electrons are fed into the MOs as follows:

$$\text{CO} \quad (\sigma 1s)^2(\sigma^* 1s)^2(\sigma 2s)^2(\sigma^* 2s)^2(\pi 2p)^4(\sigma 2p)^2 \quad (5.18)$$

The net number of bonding electrons is six, giving a bond order of three. Like nitrogen, it has a triple bond, but four of the bonding electrons have come from the oxygen and two from the carbon atom.

The NO molecule has 15 electrons, giving the ground configuration:

$$\text{NO} \quad (\sigma 1s)^2(\sigma^* 1s)^2(\sigma 2s)^2(\sigma^* 2s)^2(\pi 2p)^4(\sigma 2p)^2(\pi^* 2p)^1 \quad (5.19)$$

In this case there are five net bonding electrons, resulting in a bond order of $2\frac{1}{2}$.

### Worked Problem 5.4

**Q** What are the ground electron configurations of the short-lived molecules (a) CN, (b) BN, (c) NF and (d) FO?

**A** CN, BN, NF and FO have 13, 12, 16 and 17 electrons, respectively, and their ground electron configurations are:

| | | Bond order |
|---|---|---|
| (a) CN | $(\sigma 1s)^2(\sigma^* 1s)^2(\sigma 2s)^2(\sigma^* 2s)^2(\pi 2p)^4(\sigma 2p)^1$ | $2\frac{1}{2}$ |
| (b) BN[a] | $(\sigma 1s)^2(\sigma^* 1s)^2(\sigma 2s)^2(\sigma^* 2s)^2(\pi 2p)^4$ | 2 |
| (c) NF | $(\sigma 1s)^2(\sigma^* 1s)^2(\sigma 2s)^2(\sigma^* 2s)^2(\pi 2p)^4(\sigma 2p)^2(\pi^* 2p)^2$ | 2 |
| (d) FO | $(\sigma 1s)^2(\sigma^* 1s)^2(\sigma 2s)^2(\sigma^* 2s)^2(\pi 2p)^4(\sigma 2p)^2(\pi^* 2p)^3$ | $1\frac{1}{2}$ |

[a] Experimental evidence suggests that the configuration ... $(\pi 2p)^3(\sigma 2p)^1$ is of lower energy and is therefore the ground configuration. This is just one example of the dangers of using, for heteronuclear diatomics, MOs which have been derived for homonuclear diatomics.

For heteronuclear diatomic molecules in which both atoms are very dissimilar, no useful generalizations can be made regarding the construction of their MOs, except that the two rules which we have used for all molecules so far are still valid. These rules are that the AOs used in the LCAO method of constructing MOs must have the same symmetry with respect to the internuclear axis, and that they must have similar energies.

In HCl, for example, the hydrogen and chlorine atoms have the ground AO configurations $1s^1$ and $[Ne]3s^2 3p^5$, respectively. The ionization energy (see equation 2.14) of H is 13.598 eV while that of Cl, for the removal of a 3p electron, is 12.967 eV. Therefore these two AOs are of comparable energies. The ionization energies for the 3s, 2p, 2s and 1s AOs of Cl are very much higher, and these AOs take no part in the formation of MOs. The electrons in them remain in AOs that are almost unchanged in the molecule, and are referred to as **core electrons**.

When we consider the symmetry requirement for the formation of MOs we can see that only the $3p_z$ AO of Cl can combine with the 1s AO of H to form an MO, as shown in Figure 5.7. This MO is cylindrically symmetrical about the internuclear axis and is therefore a $\sigma$-type MO. Although the $3p_x$ and $3p_y$ AOs are of energy comparable to the 1s AO of H, they do not have the same symmetry and cannot combine with it to form MOs. The $3p_x$ and $3p_y$ AOs are almost unchanged in the molecule, but, when we consider their symmetry with respect to the internuclear axis, are clearly $\pi$-type MOs. The four electrons in them are **non-bonding electrons**. The ground configuration for the outer MOs of HCl is therefore $(\sigma)^2(\pi)^4$.

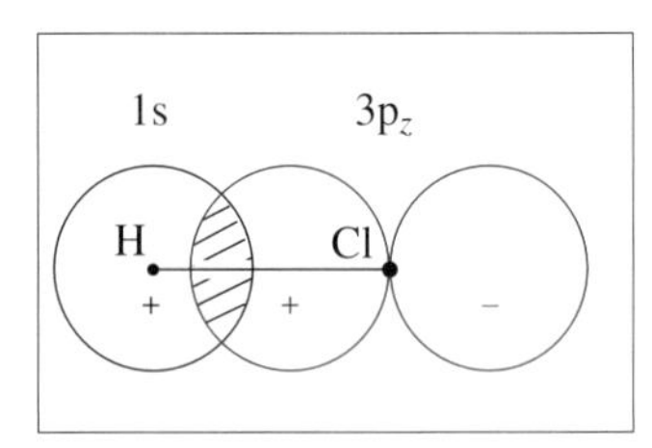


**Figure 5.7** Formation of a $\sigma$ MO in HCl

## Worked Problem 5.5

**Q** Using the representations of the AOs in Figure 3.6, and taking the internuclear axis to be the $z$-axis, use only the symmetry requirement for MO formation to show which 3d AO can overlap with the $2p_x$ AO to form an MO in a heteronuclear diatomic molecule. Which 3d AO can overlap with the $2p_z$ AO?

**A** The symmetry requirement for formation of MOs from AOs is that they should both have the same symmetry with respect to the internuclear axis. As Figure 5.8 shows, only the $3d_{xz}$ AO can overlap with the $2p_x$ AO. Only the $3d_{z^2}$ AO can overlap with the $2p_z$ AO.

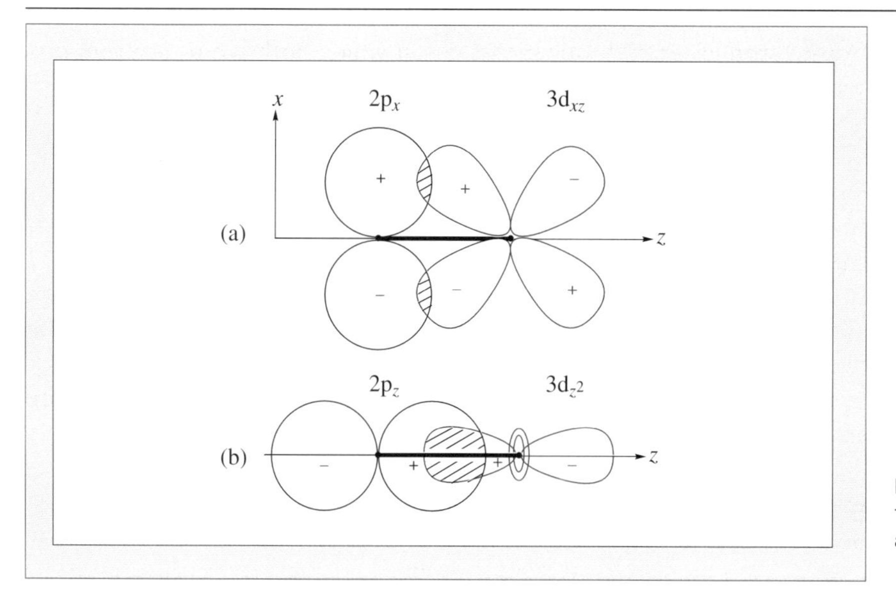


**Figure 5.8** Formation of MOs from (a) $2p_x$ and $3d_{xz}$ and (b) $2p_z$ and $3d_{z^2}$ AOs

### 5.2.2 Ground and Excited Electronic States

The method for deriving electronic states from MO configurations is very similar to that used for homonuclear diatomic molecules. The main difference is that the g or u symmetry is lost.

In the ground configuration of CO, in equation (5.18), all occupied MOs are filled, and the ground state is therefore $^1\Sigma^+$. The partially occupied MOs in the first excited configuration are $(\sigma 2p)^1(\pi^*2p)^1$. The two unpaired electrons give rise to singlet and triplet states. The only contributor to the orbital angular momentum is the $\pi^*2p$ electron. The excited states arising are therefore $^1\Pi$ and $^3\Pi$.

In the ground configuration of NO, in equation (5.19), the single electron in the $\pi^*2p$ MO gives rise to a $^2\Pi$ ground state. If this electron is promoted to the $\sigma^*2p$ MO, the resulting excited state is $^2\Sigma^+$.

**Worked Problem 5.6**

**Q** What excited states arise from the outer MO configuration $(\sigma 2p)^1(\pi^*2p)^2$ of NO?

**A** This example is not so straightforward. We start by considering the two $\pi^*2p$ electrons. Using the result from equation (5.12) we see that, so far as orbital angular momentum is concerned, they give rise to $\Sigma^+ + \Sigma^- + \Delta$ states. This result is not affected by the inclusion of the $\sigma 2p$ electron because it does not have any angular momentum.

Then we consider the spins of the three electrons in unfilled MOs. These may be all parallel, giving $S = \frac{3}{2}$ and quartet states, or two of them may be antiparallel, giving $S = \frac{1}{2}$ and doublet states. Therefore the excited states arising are $^2\Sigma^+$, $^2\Sigma^-$, $^2\Delta$, $^4\Sigma^+$, $^4\Sigma^-$ and $^4\Delta$.

In the first excited configuration of HCl, an electron is promoted from a non-bonding π MO to the σ* MO which is antibonding between the two atoms, to give the outer MO configuration $(\pi)^3(\sigma^*)^1$. The vacancy in the π orbital behaves like a single electron so that the states arising are $^1\Pi$ and $^3\Pi$.

## 5.3 Polyatomic Molecules

### 5.3.1 Molecules Containing a Carbon–Carbon Multiple Bond

Linear polyatomic molecules are the simplest in respect of their electronic structure. The reason for this is their similarity, particularly so far as their symmetry is concerned, to diatomic molecules.

Ethyne (acetylene), H–C≡C–H, is linear and has a carbon–carbon triple bond. A useful way of visualizing the makeup of the MOs concerned is to consider each carbon atom as having two sp **hybrid orbitals** at 180° to each other. These overlap with the 1s AOs of the hydrogen atoms to form σ-type MOs, as shown in Figure 5.9. Two electrons fed into each of these comprise the C–H single bonds. The other two sp hybrid orbitals, one on each carbon atom and each containing one electron, overlap to form the σ component of the triple bond. This leaves each carbon with $2p_x$ and $2p_y$ AOs which overlap to form the two π-type MOs. The one shown in the figure is $\pi 2p_x$; the $\pi 2p_y$ MO is at 90° to the one shown. Two electrons are fed into each of the π MOs to form the other two components of the triple bond.

A hybrid MO is formed by combining AOs. The combination of an s and a p AO results in two sp hybrid MOs. Combination of an s AO with two or three p AOs results in three $sp^2$ or four $sp^3$ hybrid MOs, respectively. Examples of molecules in which hybrid orbitals are useful in forming a picture of MOs are ethyne (sp), ethene ($sp^2$) and methane ($sp^3$).

In this simple treatment of MOs the LCAO method has been

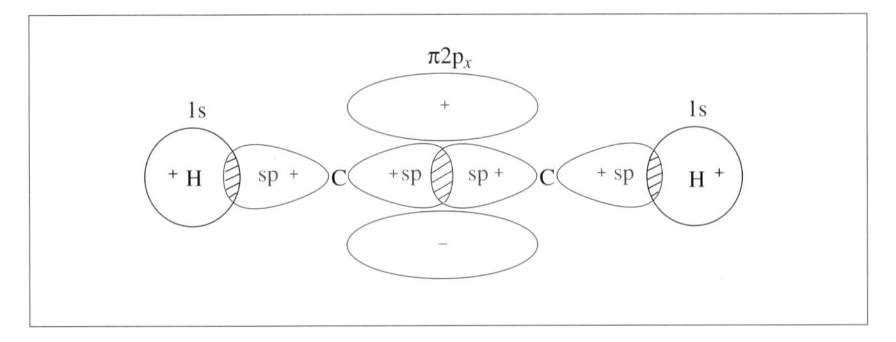


**Figure 5.9** Formation of MOs in ethyne (acetylene)

employed, as in diatomic molecules, to form the MOs from the atomic orbitals (or hybrid orbitals) on each of the atoms.

In the ground configuration of ethyne, all the electrons are in filled MOs and the ground state is $^1\Sigma_g^+$, the g subscript implying symmetric behaviour with respect to inversion through the centre of the molecule.

If, say, a fluorine atom is substituted for one of the hydrogen atoms to give fluoroethyne (fluoroacetylene), F–C≡C–H, the centre of symmetry is lost and the ground state becomes $^1\Sigma^+$.

The lowest excited MO in ethyne and fluoroethyne is a π* type. This is antibonding and similar to such MOs in diatomic molecules. As in a homonuclear diatomic molecule, promotion of an electron from a π to a π* MO results in six excited states. These are $^{1,3}\Sigma_u^+$, $^{1,3}\Sigma_u^-$ and $^{1,3}\Delta_u$ for ethyne and $^{1,3}\Sigma^+$, $^{1,3}\Sigma^-$ and $^{1,3}\Delta$ for fluoroethyne [see Worked Problem 5.3 for the $C_2$ molecule in the .... $(\pi_u 2p)^3(\pi_g 2p)^1$ excited configuration in Section 5.1.2].

Propyne, $CH_3$–C≡C–H, presents greater problems in assigning labels to electronic states. Not only is there no centre of symmetry but the cylindrical symmetry has been lost also. The symmetry classification of the molecule, and consequent labelling of electronic states, is beyond the scope of this book. However, this molecule does behave electronically in a rather similar way to, say, fluoroethyne, and it is still useful to think of the C–H and C–$CH_3$ bonds as comprising pairs of electrons in σ-type MOs and the C≡C bond as consisting of one σ- and two π-type components.

There is a similar situation in ethene (ethylene), $H_2C{=}CH_2$. Again, the symmetry properties of the skeleton of the molecule are important but will not be considered here. The overall MO picture of ethene is of four σ-type C–H bonds and a σ-type component of the C=C bond. The second component of the double bond is a π-type MO. This, and the corresponding antibonding π*-type MO, are illustrated in Figure 5.10. In the ground state the σ- and π-type MOs are filled, giving a singlet state. The lowest energy excited states result from the promotion of an electron from the π to the π* MO, giving a singlet and a triplet state.

Note that, in ethene, the loss of the cylindrical symmetry of a linear molecule, such as ethyne, results in π orbitals no longer being doubly degenerate.

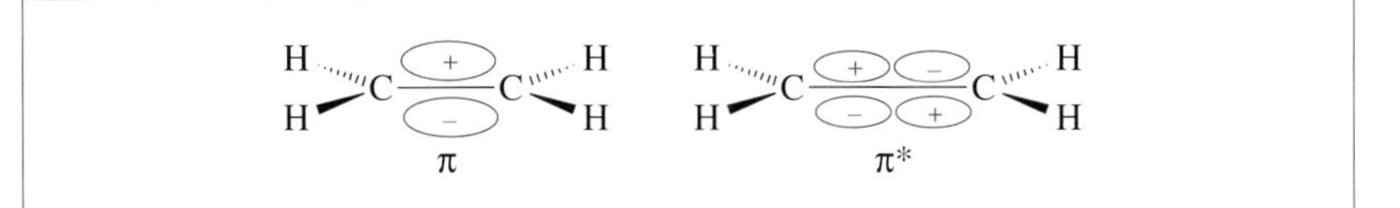


**Figure 5.10** π and π* MOs in ethene (ethylene)

It is a useful shorthand to refer to such states as ππ* states. Similarly, when π, π*, σ or σ* MOs are involved, we can refer to σπ* or πσ* excited states. However, because electrons in σ-type MOs are more tightly bound than in π-type MOs, such excited states usually lie at relatively high energies.

Another useful shorthand notation is to refer to the lowest, second lowest, *etc.*, singlet excited states as $S_1$, $S_2$, *etc.*, and the lowest, second lowest, *etc.*, triplet excited states as $T_1$, $T_2$, *etc.* This type of notation is particularly useful for closed shell molecules, *i.e.* those having a ground configuration in which all the occupied MOs are filled. The ground state is labelled $S_0$, and $T_1$ is invariably lower in energy than $S_1$.

If the ethynyl or ethenyl group is attached to an unsaturated group, as in, say, $CH_3C{\equiv}C{-}H$, $CH_3CH_2C{\equiv}C{-}H$, $CH_3CH{=}CH_2$ or $CH_3CH_2CH{=}CH_2$, only the localized MOs of the C≡C or C=C group need be considered when we are concerned only with the ground or low-lying excited electronic states. This is because the electrons in the σ-type MOs in the rest of the molecule are sufficiently tightly bound for them to play only a minor part.

### 5.3.2 Molecules Containing a Carbonyl Group

We consider here only the ground and lowest lying excited states of molecules which contain a C=O group but do not contain any other π-electron-containing group. Examples are methanal (formaldehyde), $H_2C{=}O$, ethanal (acetaldehyde), $CH_3CHO$, and propanone (acetone), $(CH_3)_2C{=}O$. For these purposes the MOs localized in the C=O group can be considered in isolation.

The σ component of the double bond in methanal is formed by overlap of an $sp^3$ hybrid orbital on carbon and a 2p orbital on oxygen. This leaves one electron in a 2p orbital on C and three electrons in 2p orbitals on O. Figure 5.11 shows the overlap of 2p orbitals to form the π component of the double bond. In the ground state, this π MO contains two electrons. The remaining two electrons are in an MO which is virtually unchanged from a 2p AO on oxygen. These form a non-bonding pair, and such an orbital is often labelled an "n" MO to indicate this non-bonding character.

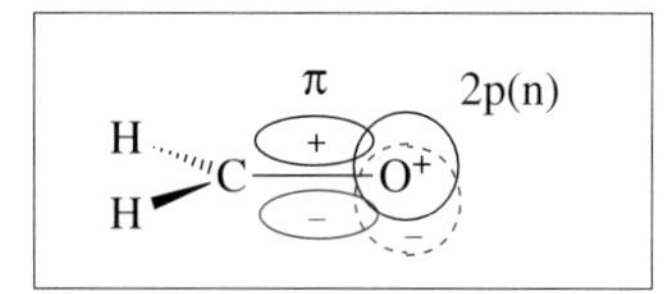


**Figure 5.11** MOs of methanal (formaldehyde)

The ground state, $S_0$, is clearly a singlet state with all the electrons in filled orbitals. The lowest energy excited MO is a π* MO, similar to that shown in Figure 5.10. The lowest energy excited states result from the promotion of a non-bonding, n, electron into the π* MO to give a triplet state, $T_1$, the lowest energy excited state, and a singlet state, $S_1$. These states are referred to as nπ* states. The $T_2$ and $S_2$ ππ* states are analogous to those of ethene, and are at higher energies.

### 5.3.3 Molecules Containing Conjugated π-Electron Systems

A conjugated π-electron molecule contains alternating single and multiple bonds. We shall consider only alternating single and double bonds. The simplest example is buta-1,3-diene, $CH_2$=CH–CH=$CH_2$. The lower energy MOs are, again, the π-type MOs. For our present purposes we can regard the carbon chain as linear, although the CCC angles are approximately 120°. An LCAO treatment of the four 2p AOs, one on each carbon atom, in a plane normal to the plane of the carbon atoms, produces four MOs. These are delocalized over the carbon chain. The three MOs of lowest energies, $\pi_1$, $\pi_2$ and $\pi_3$, are illustrated in Figure 5.12. Because the $\pi_2$ MO has one nodal plane perpendicular to the carbon chain, it is at higher energy than $\pi_1$. The $\pi_3$ has two and the $\pi_4$ MO has three such nodal planes, and have successively higher energies.

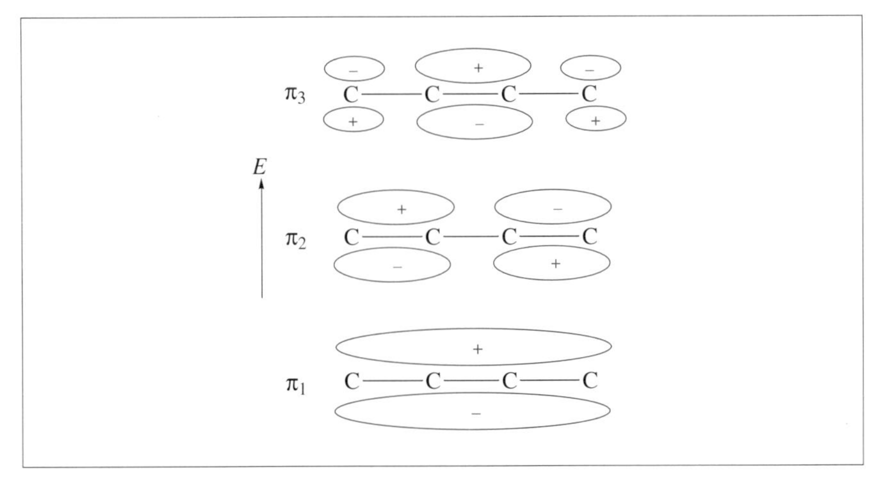


**Figure 5.12** Lower energy π MOs of buta-1,3-diene

In the ground configuration the four 2p electrons are in the $\pi_1$ and $\pi_2$ MOs, giving the outer MO configuration $(\pi_1)^2(\pi_2)^2$. The ground state, $S_0$, is a singlet state. The lowest energy excited states are ππ* states, $S_1$ and $T_1$. These result from promotion of an electron from the $\pi_2$ to the $\pi_3$ MO to give the outer MO configuration $(\pi_1)^2(\pi_2)^1(\pi_3)^1$.

All the $\pi_2$, $\pi_3$ and $\pi_4$ MOs have nodal planes perpendicular to the plane of the molecule and are, to differing extents, antibonding. It is because the $\pi_3$ MO is *more* antibonding than $\pi_2$ that we can usefully label it π*.

One of the most important conjugated π-electron systems is that in the benzene molecule. Once the skeleton of six σ-type C–C bonds and the six C–H bonds has been taken account of, there remain six electrons in 2p AOs perpendicular to the carbon ring. LCAO treatment of these results in six MOs, $\pi_1$ to $\pi_6$, with 0 to 3 nodal planes perpendicular to the ring. These are shown in Figure 5.13. The lowest and highest energy MOs, $\pi_1$ and $\pi_6$, are non-degenerate, but the $\pi_2$,$\pi_3$ and $\pi_4$,$\pi_5$ MOs form doubly degenerate pairs.

In the ground MO configuration the six π electrons are accommodated to give the outer MO configuration $(\pi_1)^2(\pi_{2,3})^4$ and a singlet ground state $S_0$. The lowest energy excited configuration is one in which an electron is promoted from $\pi_{2,3}$ to $\pi_{4,5}$ to give $(\pi_1)^2(\pi_{2,3})^3(\pi_{4,5})^1$. The vacancy in the $\pi_{2,3}$ MO behaves like a single electron, so far as deriving the resulting electronic states is concerned. Similar to the case of two electrons in

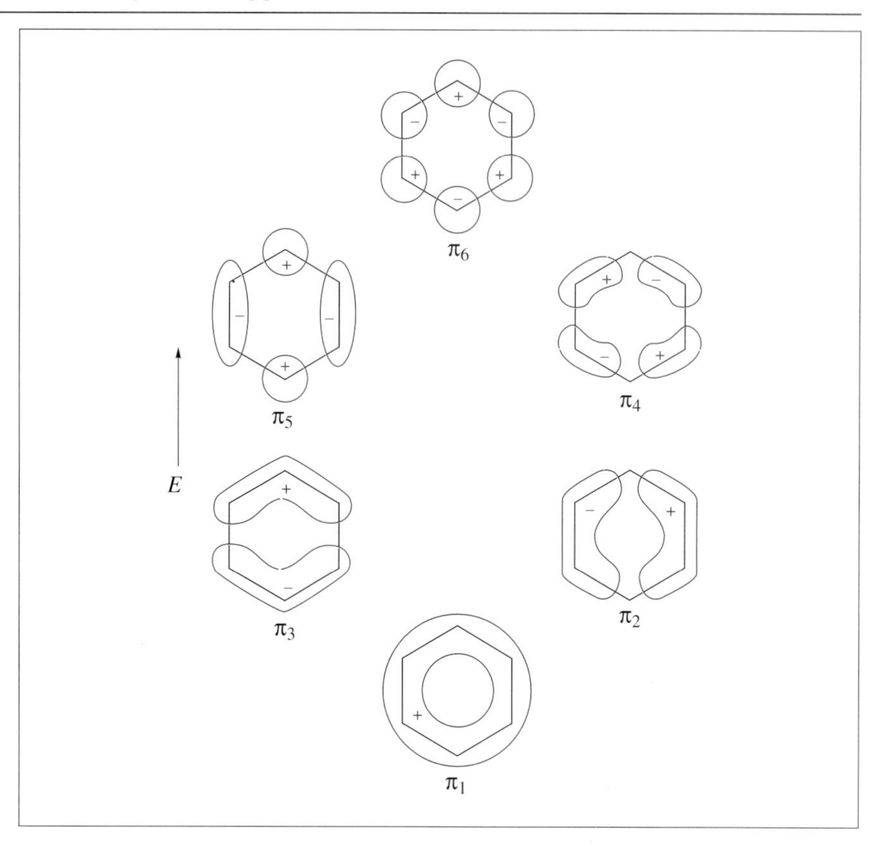


**Figure 5.13** π MOs of benzene

different π MOs in a diatomic molecule (see, for example, equation 5.15 *et seq.*), three singlet ($S_1$, $S_2$ and $S_3$) and three triplet ($T_1$, $T_2$ and $T_3$) excited states result. All are ππ* states, the $\pi_{4,5}$ being more antibonding than the $\pi_{2,3}$ MOs.

In the pyridine (azabenzene) molecule, $C_5H_5N$, one of the CH groups of benzene has been replaced by a nitrogen atom. The orbitals on the nitrogen atom can be regarded as comprising three $sp^2$ hybrid MOs, two of which are bonded to adjacent carbon atoms, leaving a 2p orbital perpendicular to the ring and taking part in π MOs which are similar to those of benzene. The remaining $sp^2$ orbital is in the plane of the ring and takes two electrons. The orbital is a non-bonding, n, orbital which is the highest energy occupied orbital, sometimes labelled with the acronym HOMO (highest occupied molecular orbital). Promotion to the lowest energy unoccupied orbital (LUMO) results in a singlet and a triplet nπ* state, $S_1$ and $T_1$, the lowest energy excited states. There are also ππ* singlet and triplet excited states at higher energies. These are similar to ππ* states of benzene, except that the lower symmetry of pyridine results in the loss of the double degeneracies of the π MOs in Figure 5.13.

## Summary of Key Points

**1.** *Method of linear combination of atomic orbitals*
In the vicinity of one of the atoms in a diatomic molecule, a molecular orbital resembles the corresponding atomic orbital. Two MOs, one bonding and one antibonding, always result from the LCAO treatment of two AOs.

**2.** *Overlap of molecular orbitals*
Positive or negative overlap results in bonding or antibonding character, respectively.

**3.** *Aufbau principle applied to all molecules*
Available electrons are fed into the MOs, in order of increasing energy, to give the ground configuration.

**4.** *Use of Russell–Saunders coupling approximation for electron spin and orbital angular momenta*
Introduction of the quantum numbers $S$, $\Lambda$ and $\Sigma$ for diatomic molecules.

**5.** *Derivation of ground and excited electronic states from MO configurations of diatomic molecules*
Use of g and u symmetry in homonuclear diatomic molecules. Special cases of two electrons in either the same or different $\pi$ MOs.

**6.** *Heteronuclear diatomic molecules comprising similar or very different atoms*
For the case of similar atoms, MOs which resemble those for homonuclear diatomic molecules can be used. When the atoms are very different, each molecule must be treated individually.

**7.** *Molecular orbitals and electronic states of polyatomic molecules*
Linear polyatomic molecules treated in a similar way to diatomic molecules. Treatment of a local group, such as C=O, is also similar to that of a diatomic molecule. MOs in various $\pi$-electron containing molecules such as ethyne, ethene, buta-1,3-diene and benzene. Labelling of electronic states as, for example, $\pi\pi^*$, $n\pi^*$ or $\sigma\pi^*$ states, and as S or T states, depending on whether they are singlet or triplet, respectively.

## Problems

**5.1.** What is the bond order for the following excited configurations:
(a) $N_2$ $[Be_2](\pi_u 2p)^3(\sigma_g 2p)^2(\pi_g 2p)^1$
(b) $O_2$ $[Be_2](\sigma_g 2p)^2(\pi_u 2p)^3(\pi_g 2p)^3$
(c) $O_2^+$ $[Be_2](\sigma_g 2p)^2(\pi_u 2p)^3(\pi_g 2p)^2$

**5.2.** Discuss the general principles involved in comparing the MOs of $S_2$ and SO with those of $O_2$.

**5.3.** In a diatomic molecule, such as $I_2$, which contains at least one heavy atom, coupling between the total orbital and total spin angular momenta (spin–orbit, or *LS*) coupling may be large. Then, the Hund's case (c), rather than Hund's case (a), coupling approximation is appropriate [Hund's case (a) coupling has been discussed in Section 5.1.2]. Draw a vector diagram representing case (c) coupling. What are the "good" quantum numbers for this coupling case?

**5.4.** In symbolic form, what is the ground state of (a) $NO^+$ and (b) $NO^-$?

**5.5.** Describe the main differences between MO descriptions of the ground electronic states of (a) hydrogen cyanide (HC≡N) and ethyne (HC≡CH), and (b) pyrazine (1,4-diazabenzene) and pyridine (azabenzene).

# 6 Molecular Vibrations

**Aims**

In this chapter the reader will be introduced to vibrational motion, and its quantized nature, in diatomic and polyatomic molecules. By the end of this chapter you should be able to:

- Understand the treatment of the vibration of a diatomic molecule in the harmonic oscillator approximation
- Understand the effects of the vibration being anharmonic
- Calculate the number of normal vibrations in a polyatomic molecule
- Recognize some of the types of vibration that a polyatomic molecule may have
- Understand what is meant by a group vibration

## 6.1 Introduction

The fact that in this chapter we are able to discuss vibrational motion independently of the motion of the electrons, which was discussed in Chapter 5, is due to the general usefulness of the **Born–Oppenheimer approximation**. The validity of this approximation follows from the fact that the nuclei in a molecule are very much heavier than the electrons and, therefore, move much more slowly. An important result of this is that the total wave function, $\psi_{total}$, can be factorized into electronic and nuclear contributions, $\psi_e$ and $\psi_n$, respectively:

$$\psi_{total} = \psi_e \psi_n \tag{6.1}$$

A further approximation involves the factorization of $\psi_n$ into vibrational and rotational contributions, $\psi_v$ and $\psi_r$, respectively, giving:

$$\psi_{\text{total}} = \psi_e \psi_v \psi_r \tag{6.2}$$

This factorization results in the total energy of a molecule being the sum of the contributions from the electrons, the vibrations and the rotations:

$$E_{\text{total}} = E_e + E_v + E_r \tag{6.3}$$

The approximations which result in the simplicity of this equation are not always valid, but such cases will not concern us in this book.

Molecules are held together by electrons in bonding molecular orbitals, the nature of which has been discussed in Chapter 5. Although the representation of bonds by straight lines between atoms tends to give the impression that the bonds behave like rigid rods, this is not so. They are much better represented, for spectroscopic purposes, by springs connecting the atoms. This is illustrated, for a diatomic molecule (homonuclear or heteronuclear), in Figure 6.1(a).

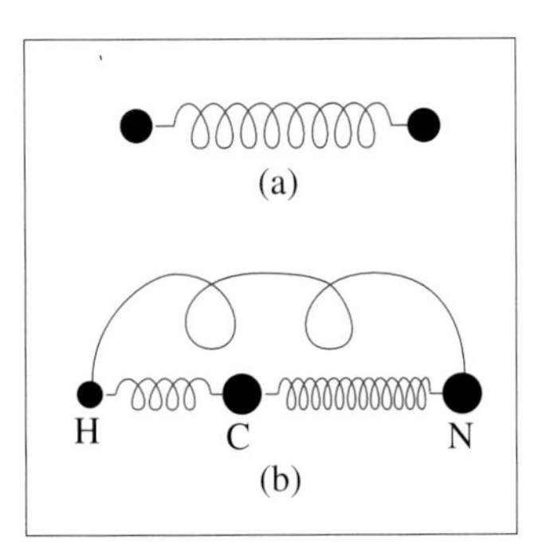


**Figure 6.1** Ball-and-spring model for vibrations of (a) a diatomic molecule and (b) H–C≡N

The strength of the spring varies from one molecule to another; in particular, it increases with the multiplicity of the bond. Clearly, the spring is stronger for a triple bond, such as that in $N_2$, than for the double bond in $O_2$. For a single bond, as in $F_2$ and HCl, the spring is weaker still.

For a linear polyatomic molecule such as HCN (Figure 6.1b) there are springs representing each bond, a strong one for the triple bond and a much weaker one for the single bond. A further complication is that the molecule can also bend about the carbon atom. It is principally the hydrogen atom which is involved in this motion, because it is much lighter than the nitrogen atom. This bending motion is represented by a further, very weak spring.

From our consideration of just these two types of molecule we have encountered two types of **vibrational mode**. The first is a **stretching** mode, as in the diatomic example and in the C–H and C≡N stretching modes in HCN. The second type is the **bending** mode of vibration in HCN.

## 6.2 Vibration in a Diatomic Molecule

### 6.2.1 Treated as a Harmonic Oscillator

The vibration in a diatomic molecule can be treated as behaving like the stretching and compression of a spring. Stretching or compressing a spring produces **potential energy**. When the spring is released, this energy is translated into a **restoring force** which results in vibration about the equilibrium length of the spring. Similarly, when a bond is stretched or compressed, and then released, vibrational motion occurs.

If the vibration involves only small displacements of the bond length,

$r$, from the equilibrium bond length, $r_e$, the restoring force, $f$, is proportional to the displacement, $x$, of the bond length from its equilibrium value, where:

$$x = r - r_e \quad (6.4)$$

This law of proportionality is known as **Hooke's law**:

$$f = -kx \quad (6.5)$$

This law is also obeyed for small displacements of a so-called Hooke's law spring. Any oscillator, such as a molecule undergoing vibrational motion which obeys Hooke's law, is a **harmonic oscillator**. The quantity $k$ in equation (6.5) is the **force constant**, and is large for strong bonds (springs) and small for weak bonds. This is illustrated by the values for $k$ of 328.6, 1177 and 2296 N m$^{-1}$ for $Cl_2$, $O_2$ and $N_2$, having a single, double and triple bond, respectively.

The SI unit of force is the newton for which the symbol is N, where 1 N = 1 m kg s$^{-1}$.

The potential energy, $V(x)$, resulting from the displacement $x$ is given by:

$$V(x) = \tfrac{1}{2}kx^2 \quad (6.6)$$

The energy associated with vibration is quantized: like the electronic energy, it can only take specific values. These discrete vibrational energy levels are obtained by solution of the Schrödinger equation of equation (3.19), which also gives the vibrational wave functions, $\psi_v$. The resulting vibrational energy levels, $E_v$, are given by:

$$E_v = h\nu(v + \tfrac{1}{2}) \quad (6.7)$$

where v is the vibrational quantum number, which can take any integral value 0, 1, 2, 3, ..., and $\nu$ is the vibration frequency given by:

$$\nu = (1/2\pi)(k/\mu)^{\frac{1}{2}} \quad (6.8)$$

The quantity $\mu$ is the reduced mass, and is given by:

$$\mu = m_1m_2/(m_1 + m_2) \quad (6.9)$$

where $m_1$ and $m_2$ are the masses of atoms 1 and 2.

The unit which is used for atomic mass is the **atomic mass unit**, for which the symbol is "u". This unit is defined in terms of the mass of an atom ($m_a$) of the $^{12}C$ isotope of carbon, which is taken to be exactly 12 u, so that:

$$1\ \text{u} = m_a(^{12}\text{C})/12 \tag{6.10}$$

This equation shows that 1 u is the mass of one atom of $^{12}$C, for which the best value we have is:

$$1\ \text{u} = 1.660\ 538\ 73 \times 10^{-27}\ \text{kg} \tag{6.11}$$

The accuracy to which this value is known is limited by the accuracy to which the Avogadro constant, $N_A$, is known.

### Worked Problem 6.1

**Q** Calculate, in kg and to five significant figures, the reduced mass of the $^{10}$B$^{1}$H molecule, given that $m(^{10}\text{B}) = 10.0129$ u and $m(^{1}\text{H}) = 1.00783$ u.

**A** From equation (6.9):

$$\begin{aligned}\mu &= m(^{10}\text{B})m(^{1}\text{H})/[m(^{10}\text{B}) + m(^{1}\text{H})]\\ &= 10.0129 \times 1.00783/(10.0129 + 1.00783)\ \text{u}\\ &= 0.915\ 665\ \text{u}\\ &= 0.915\ 665 \times 1.660\ 54 \times 10^{-27}\ \text{kg}\\ &= 1.5205 \times 10^{-27}\ \text{kg}\end{aligned}$$

Note that six figures have been retained until the last line of the calculation. This follows the general principle that an extra figure is included in each step of a calculation, until the final result is obtained.

Note that the reduced mass for $^{10}$B$^{1}$H, having one atom with much larger mass than the other, is similar to the mass of the lighter atom. This is consistent with the fact that, during vibrational motion, it is the hydrogen atom which moves much more than the boron atom.

Figure 6.2 shows a plot of the potential energy, $V(r)$, against the internuclear separation, $r$. Equations (6.6) and (6.4) show that the relationship produces a parabola with the minimum potential energy at the equilibrium internuclear separation, $r_e$. Also shown are the vibrational energy levels with $v = 0$–3. A very important aspect of these is that the lowest level, that with $v = 0$, is not at the zero of energy. According to equation (6.7), the energy of the $v = 0$ level is $\frac{1}{2}h\nu$, known as the **zero-point energy**. The fact that the vibrational energy can never be less than the zero-point energy, even when the sample is cooled to 0 K, is a consequence of the uncertainty principle of equation (3.17).

Equation (6.7) shows that, above the $v = 0$ level, the levels have equal spacings of $h\nu$. In terms of wavenumber, which is used invariably in vibrational spectroscopy, the vibrational spacing is $hc\omega$ (= $h\nu$), where $\omega$ is the vibration wavenumber and $c$ is the speed of light. In terms of

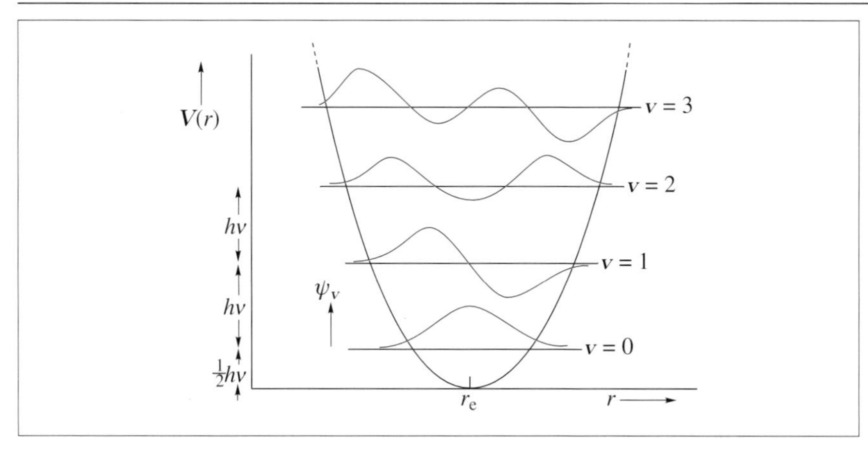


**Figure 6.2** Potential energy curve, and a few vibrational levels and wave functions, for a diatomic molecule acting as a harmonic oscillator

wavenumber, the energy levels, or **vibrational term values** as they are now called, are given by:

Although the distinction between vibration frequency and vibration wavenumber is clear, it is unfortunately the case that the quantity $\omega$ is often referred to as the vibration frequency.

$$G(v) = \omega(v + \tfrac{1}{2}) \tag{6.12}$$

Solution of the Schrödinger equation for the harmonic oscillator also gives the vibrational wave functions, $\psi_v$. These are shown for the $v$ = 0–3 levels in Figure 6.2.

### 6.2.2 Treated as an Anharmonic Oscillator

When a spring is compressed or stretched beyond certain limits, Hooke's law is no longer obeyed. Similarly, for large displacements the stretching and contraction of the bond of a diatomic molecule do not obey Hooke's law. The vibration no longer behaves like that of a harmonic oscillator but like that of an **anharmonic oscillator**.

Figure 6.3 shows how the potential energy curve is changed from that for the harmonic oscillator in Figure 6.2. At shorter bond lengths, little further shortening occurs when the potential energy of compression is increased: the curve rises steeply. At longer bond lengths, the bond begins to weaken: the force constant decreases. Eventually, when it is stretched even further, the bond breaks, and **dissociation** occurs.

The vibrational energy levels within an anharmonic potential curve are no longer equally spaced above the zero-point level. As shown in Figure 6.3, they converge smoothly towards the dissociation limit. It is important to remember that, for a molecule in the ground electronic state, dissociation always produces **neutral atoms**, even in a such a highly polar molecule as HF.

The smooth convergence of the energy levels is taken into account by adding successively smaller terms to the term value expression of equation (6.12) to give:

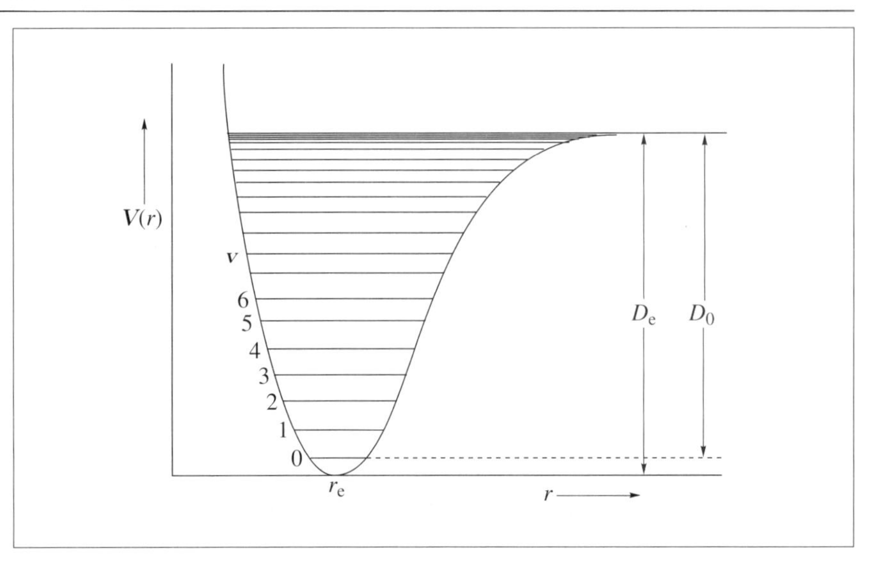


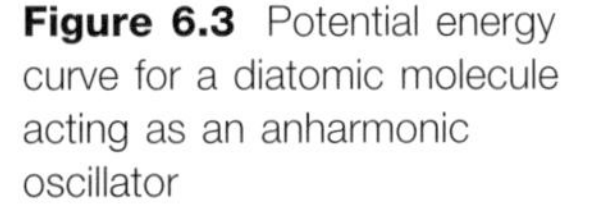

**Figure 6.3** Potential energy curve for a diatomic molecule acting as an anharmonic oscillator

$$G(v) = \omega_e(v + \tfrac{1}{2}) - x_e(v + \tfrac{1}{2})^2 + y_e(v + \tfrac{1}{2})^3 + \ldots \qquad (6.13)$$

The quantities $x_e$, $y_e$, ... are **anharmonic constants**. The negative sign is placed in front of the second term for convenience; this ensures that all values of $x_e$ are positive.

The anharmonic constants for a diatomic molecule have usually been represented, for historical reasons, by $\omega_e x_e$, $\omega_e y_e$, ..., rather than by $x_e$, $y_e$, ... . Because the former symbolism can be confusing and, more importantly, cannot be used for polyatomic molecules, we are using the simpler symbolism here.

## Worked Problem 6.2

**Q** Using the given data, calculate the $v$ = 0, 1 and 2 vibrational term values for (a) $^1H_2$, for which $\omega_e$ = 4401.2 cm$^{-1}$, $x_e$ = 121.3 cm$^{-1}$ and $y_e$ = 0.8 cm$^{-1}$, and (b) $^{79}Br_2$, for which $\omega_e$ = 325.32 cm$^{-1}$, $x_e$ = 1.08 cm$^{-1}$ and $y_e$ = –0.002 cm$^{-1}$. For each molecule, compare the $v$ = 1–0 and 2–1 separations.

**A** Using equation (6.13) gives:

| | $G(0)$/cm$^{-1}$ | $G(1)$/cm$^{-1}$ | $G(2)$/cm$^{-1}$ | [$G(1)$–$G(0)$] /cm$^{-1}$ | [$G(2)$–$G(1)$] /cm$^{-1}$ |
|---|---|---|---|---|---|
| $^1H_2$ | 2170.4 | 6331.6 | 10257.4 | 4161.2 | 3925.8 |
| $^{79}Br_2$ | 162.39 | 485.54 | 806.52 | 323.15 | 320.98 |

Because it is such a light molecule, $^1H_2$ has a very much larger fundamental vibration wavenumber than $^{79}Br_2$. Consequently, the vibrational energy levels lie much higher up the vibrational potential function, and the effects of anharmonicity are much greater. This can be seen by comparing the closing up of the first few vibrational levels: this is much more pronounced for $^1H_2$ than for $^{79}Br_2$.

Figure 6.3 shows that the dissociation energy can be measured relative to that at equilibrium, giving $D_e$, or at the zero-point level, giving $D_0$. Since a molecule can never be at equilibrium, only $D_0$ may be obtained directly. $D_e$ can be obtained from it by adding the zero-point energy.

There have been various attempts to modify the expression for the potential energy of a harmonic oscillator, given in equation (6.6), to take account of anharmonic behaviour. One of the most frequently used is the **Morse potential**:

$$V(x) = D_e[1 - \exp(-ax^2)] \tag{6.14}$$

where $x = r - r_e$ and $a$ is a constant. This potential is approximate, but has the great virtue that $V(x) \rightarrow D_e$ as $x \rightarrow \infty$, *i.e.* as the bond breaks.

One possible method of obtaining the dissociation energy is through the observation of as many vibrational levels as possible. The separations, $\Delta G_{v+\frac{1}{2}}$ [$= G(v + 1) - G(v)$], of successive vibrational levels are plotted against $v + \frac{1}{2}$. Figure 6.3 shows that:

$$D_0 = \Sigma_v \Delta G_{v+\frac{1}{2}} \tag{6.15}$$

In general, the symbol $\Sigma_n$ indicates the sum of all quantities labelled by $n$.

*i.e.* $D_0$ is the sum of all the vibrational level separations. Therefore $D_0$ is given by the area under the graph, when it has been extrapolated to $\Delta G_{v+\frac{1}{2}} = 0$.

However, this is not as straightforward as it might appear. Firstly, only a limited number of vibrational levels can be observed experimentally: usually observations cease far below the dissociation limit. This means that the graph must be extrapolated to $\Delta G_{v+\frac{1}{2}} = 0$, a so-called **Birge–Sponer extrapolation**. Secondly, the graph is a straight line only if all the anharmonic constants, other than $x_e$, are zero. Consequently there may be a long extrapolation of a non-linear graph, as illustrated in Figure 6.4. This figure shows that the true curve usually bends

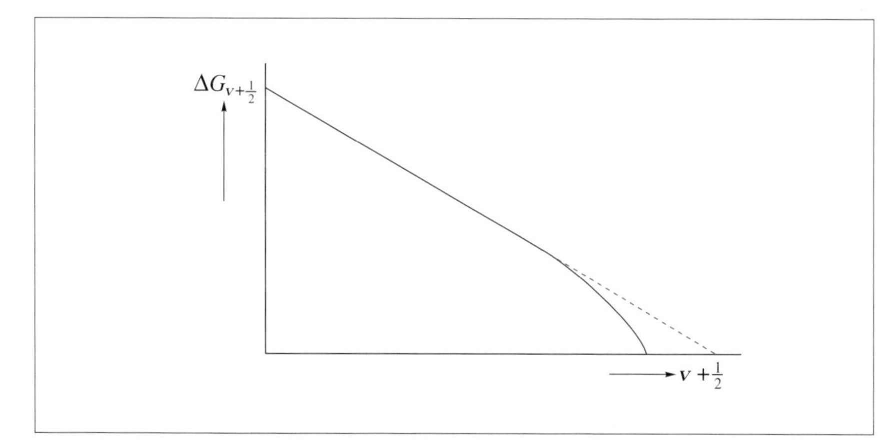


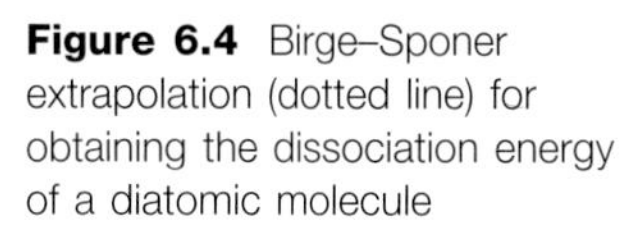
**Figure 6.4** Birge–Sponer extrapolation (dotted line) for obtaining the dissociation energy of a diatomic molecule

"downwards", so that an extrapolation from levels far below the dissociation limit will tend to overestimate the value of $D_0$.

## 6.3 Vibration in Polyatomic Molecules

A diatomic molecule has only one vibration, a stretching vibration. Figure 6.1(b) shows an example of a linear triatomic molecule having two stretching and one bending vibrations. The $H_2O$ molecule is non-linear, but also has three vibrations: these are shown in Figure 6.5. The vibrational modes, labelled $\nu_1$, $\nu_2$ and $\nu_3$, are represented by arrows attached to the atoms. The lengths of the arrows indicate the relative magnitudes of the motions of the atoms in the directions shown. In all three vibrations the movement of the oxygen atom is very small compared to that of the much lighter hydrogen atoms.

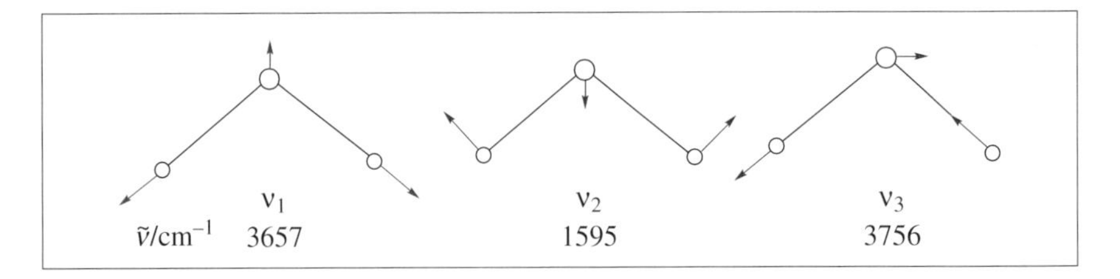


**Figure 6.5** The three normal modes of vibration of the $H_2O$ molecule

There are rules regarding the numbering of vibrations in polyatomic molecules. We shall not consider them here except to note that the bending vibration of a triatomic molecule is always labelled $\nu_2$.

The $H_2O$ molecule has two O–H bonds, each of which can stretch and contract, as in a diatomic molecule. However, because the bonds are identical, the symmetry of the situation dictates that these motions are either in-phase, as in $\nu_1$, or out-of-phase, as in $\nu_3$. These vibrations are referred to as the **symmetric** and **asymmetric stretching** vibrations, respectively. $\nu_2$ is the bending vibration.

When we consider polyatomic molecules containing more than three atoms, it is no longer obvious how many modes of vibration they have. In fact, a linear molecule has $3N - 5$ **normal modes** of vibration, and a non-linear molecule has $3N - 6$, where $N$ is the number of atoms in the molecule.

### Box 6.1 Number of Normal Modes of Vibration in a Polyatomic Molecule

These expressions for the number of normal modes of vibration in a polyatomic molecule are easily justified.

Each atom in a molecule has three degrees of freedom, which can be regarded as the freedom to move in the $x$, $y$ or $z$ direction. Therefore a molecule containing $N$ atoms has a total of $3N$ degrees of freedom. For a non-linear molecule, three of these involve translation of the whole molecule along the *x*, *y* or *z* axis. Another three involve rotation of the whole molecule about the *x*, *y* or *z* axis. None of these six degrees of freedom involves vibration, in which atoms move relative to each other. Therefore the remaining degrees of freedom, $3N - 6$, are vibrational degrees of freedom, or normal modes of vibration.

For a linear molecule, the only difference is that there are only two rotational degrees of freedom. There is no degree of freedom due to rotation about the internuclear *z*-axis; since the nuclei lie on the axis, it does not require any energy to perform such a rotation. Therefore a linear molecule has $3N - 5$ normal modes of vibration.

### Worked Problem 6.3

**Q** How many normal modes of vibration do the following molecules have: (a) prop-2-ynenitrile (cyanoethylene), (b) propanone (acetone) and (c) methylbenzene (toluene)?

**A** (a) H–C≡C–CN is a linear molecule with five atoms. Since $3N - 5 = 10$, it has ten normal vibrations.
(b) $CH_3COCH_3$ is a non-linear molecule with ten atoms. Since $3N - 6 = 24$, it has 24 normal vibrations.
(c) $C_6H_5CH_3$ is non-linear, has 15 atoms and 39 normal vibrations.

The linear molecule ethyne (acetylene) has four atoms and therefore seven normal modes, illustrated in Figure 6.6. Like $H_2O$, the molecule has two identical bonds which can stretch in-phase, $\nu_1$, or out-of-phase, $\nu_3$, the symmetric and asymmetric stretching vibrations, respectively. $\nu_2$ is the C–C stretching vibration.

The remaining four normal modes are all bending vibrations. $\nu_4$ is the *trans* and $\nu_5$ the *cis* bending vibration. Figure 6.6 shows two forms of $\nu_4$, in which the bending is in either the *xz*- or the *yz*-plane. The vibration is doubly degenerate because it requires the same energy for vibration in either plane, but the wave functions for the two forms are not the same: one is a function of *x* and *z*, while the other is a function of *y* and *z*. Similarly, $\nu_5$ is also doubly degenerate.

In Figures 6.5 and 6.6 the wavenumber of the $v = 1$–$0$ interval is given for each of the vibrations of water and ethyne. These wavenumbers illustrate some very important facts regarding vibrations in polyatomic molecules. For example, the wavenumbers of stretching vibrations

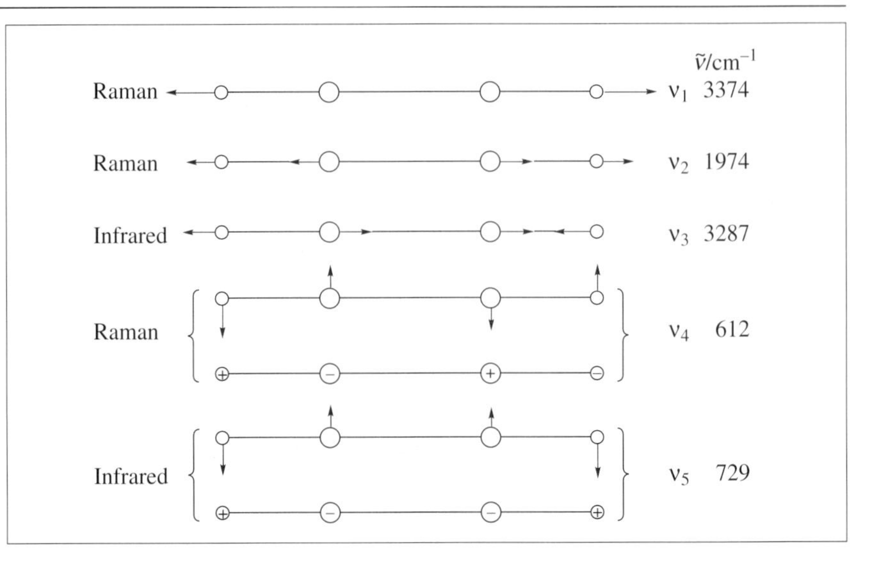


**Figure 6.6** The normal modes of vibration of ethyne (acetylene)

involving hydrogen atoms, such as $\nu_1$ and $\nu_3$ of both water and ethyne, are very high, in excess of 3000 cm$^{-1}$. The wavenumber for the stretching of a multiple bond, such as the triple bond involved in $\nu_2$ of ethyne, is much lower. The examples of water and ethyne also illustrate the fact that bending vibration wavenumbers are relatively low; this is particularly true in ethyne, where the fairly heavy carbon atoms are involved.

These few examples serve to introduce the very important concept of **group vibrations** in polyatomic molecules. In ethyne, for example, the strengths of the C–H and C≡C bonds are sufficiently different for the stretching vibrations to be almost independent, involving either the C–H bonds ($\nu_1$ and $\nu_3$) or the C≡C bond ($\nu_2$). Because of this independence it is not surprising that the wavenumber of the stretching vibration of a C≡C bond in, for example, $CH_3C{\equiv}CH$ and $C_6H_5C{\equiv}CH$ is very similar to that in ethyne. In other words, the C≡C stretching mode is a well-behaved group vibration with a wavenumber which can be transferred, with a reasonable degree of approximation, to other molecules containing the same group. Similarly, in ethene and ethane, the C=C and C–C stretching vibration wavenumbers are 1623 and 993 cm$^{-1}$, respectively, and are useful group vibrations. The use of such group vibrations does not extend to molecules containing conjugated multiple bonds, such as the C=C–C≡C group, in which the stretching of all three bonds is not even approximately independent.

The concept of a group vibration is extremely important in the application of vibrational spectroscopy to chemical analysis. If, for example, one stage in an organic synthesis involves the introduction of a C=O group, the observation in the vibrational spectrum of a band typical of this group will be important evidence that the reaction has been successful.

The calculation of the forms of the vibrations of a polyatomic molecule, as illustrated by the arrows for the various vibrations of water and ethyne in Figures 6.5 and 6.6, is a difficult classical, rather than a quantum mechanical, problem. The methods used in these calculations will not concern us in this book.

A potential energy curve, such as that in Figure 6.3, for a diatomic molecule is two-dimensional, one dimension for the internuclear distance, $r$, and the other for the potential energy, $V(r)$. For a polyatomic molecule, with $3N - 5$ or $3N - 6$ vibrational modes, the potential energy curve is replaced by a **hypersurface** in which potential energy is plotted against all the vibrational coordinates. It is, of course, impossible to represent such a surface pictorially. Instead, we can treat each vibration in isolation, by making a two-dimensional plot of potential energy against the appropriate vibrational coordinate. Such a plot is an approximation, particularly in regions of high potential energy, and is obtained by making a two-dimensional cross-section in the multidimensional hypersurface.

The potential energy curve for any stretching vibration, such as the C–H stretch in HCN, is expected to resemble that for a diatomic molecule. Such a curve is shown in Figure 6.7(a), in which the coordinate $r$ for a diatomic molecule is replaced by the general coordinate $Q$, which is known as the **normal coordinate**. The potential energy curve is that of an anharmonic oscillator, showing the dissociation limit when the hydrogen atom is removed.

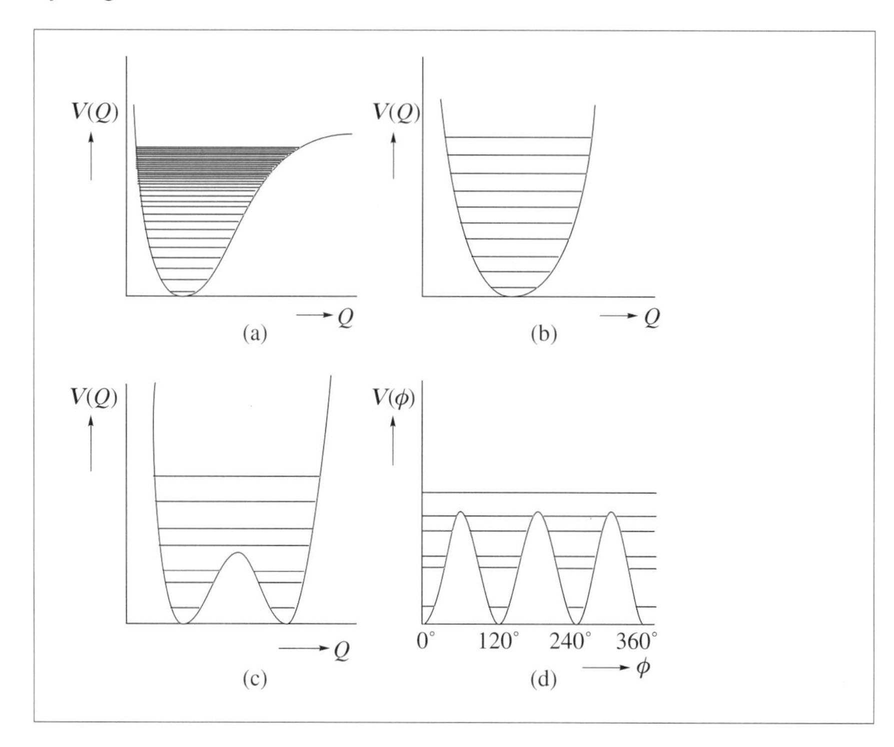


**Figure 6.7** Potential energy curves for (a) a stretching, (b) a bending, (c) an inversion and (d) a torsional vibration

However, we should not pursue this approximate way of treating stretching vibrations too far. Consider, for example, the symmetric C–H stretching vibration, $\nu_1$, of ethyne. If sufficient energy is put into this vibrational mode, does it dissociate into $C_2$ + 2H? The intuitive answer is "no", because it would require an excessive amount of energy to break both C–H bonds simultaneously. Intuition suggests that the lowest energy dissociation limit will correspond to the loss of only one hydrogen atom to give $C_2H$ + H, and this is what happens. At increasingly high energy the stretching motion becomes concentrated in one of the C–H bonds. The normal mode treatment of the vibration breaks down, and a **local mode** treatment becomes more appropriate. In this treatment the stretching of only one C–H bond is treated rather like that of a diatomic molecule. We shall not consider this local mode treatment any further, but it is particularly appropriate in molecular dynamics, in which dissociation processes in polyatomic molecules are very important.

For a bending vibration, such as $\nu_2$ of HCN (Figure 6.1b), or $\nu_4$ or $\nu_5$ of ethyne (Figure 6.6), it is clear that increased potential energy can never lead to dissociation. The corresponding potential energy curve is quite different from that in Figure 6.7(a). Because the bending motion can take place to either side of the internuclear axis and encounters increasing resistance at large amplitudes, the curve is symmetrical and steep-sided, as shown in Figure 6.7(b).

There are other types of vibration which do not lead to dissociation. One of these is the **inversion vibration**. The examples shown are of ammonia, in Figure 6.8(a), and phenylamine (aniline), in Figure 6.8(b). Both molecules are non-planar. The hydrogen atoms can move above and below the plane that the molecule would have if it were planar; in other words, the planar conformation does not correspond to an energy minimum. There are two identical energy minima corresponding to the hydrogen atoms being above or below the "plane". The planar conformation is at the top of the **energy barrier**. The resulting W-shaped potential energy curve is shown in Figure 6.7(c). Whereas the vibrational energy levels within the potentials in Figure 6.6(a) and (b) are similar to those we have encountered previously, smoothly converging to high energy in Figure 6.6(a), or more equally spaced in Figure 6.6(b), those within a W-shaped potential are very different. Energy levels below the barrier in Figure 6.7(c) are split due to quantum mechanical **tunnelling** through the barrier. The splitting increases as the top of the barrier is approached. Immediately above the barrier the energy levels are still irregular, but at higher energy they settle down to become evenly spaced.

In a classical, as opposed to a quantum mechanical, treatment of an inversion vibration, the barrier must be surmounted in order to go from one minimum in the potential to the other. In a quantum mechanical treatment, tunnelling may occur below the top of the barrier. This tunnelling takes time to occur. The narrower the barrier, and/or the lighter the tunnelling atoms, the faster is the tunnelling and the larger the splitting of energy levels. For example, tunnelling is much slower, and the splittings much smaller, in $N^2H_3$ than in $N^1H_3$.

A **torsional vibration** involves the internal twisting, clockwise or anticlockwise, of one group of atoms in the molecule relative to another. Figure 6.8(c) shows the example of the torsional vibration in 1-fluoro-2-methylbenzene (2-fluorotoluene) in which the $CH_3$ group is twisting relative to the benzene ring. In this case, the potential energy repeats for every rotation by an angle $\phi$ of 120°, and there is a three-fold energy barrier to the torsional motion. In this molecule the minima in the potential correspond to the so-called *pseudo-trans* conformation in which a hydrogen atom of the $CH_3$ group is *trans* to the fluorine atom. The potential energy curve is shown in Figure 6.7(d).

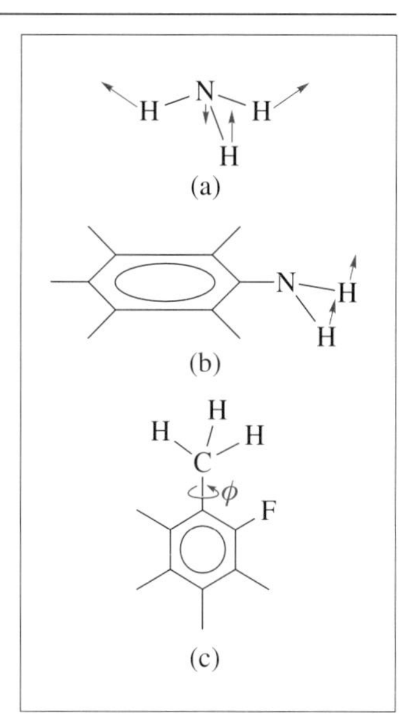


**Figure 6.8** The inversion vibration of (a) ammonia and (b) phenylamine (aniline), and (c) the torsional vibration of 1-fluoro-2-methylbenzene (2-fluorotoluene)

### Worked Problem 6.4

**Q** Assuming an *n*-fold torsional barrier, what is the value of *n* for the torsional motion in (a) methylbenzene (toluene), (b) 1-methyl-4-fluorobenzene (4-fluorotoluene), (c) methanal (acetaldehyde) and (d) phenol?

**A** (a) $n = 6$, because the torsional vibrational energy repeats every 60°. (b) $n = 6$ [the fluorine atom in the (symmetrical) 4-position does not affect the nature of the barrier]. (c) $n = 3$ (similar to the example of 1-fluoro-2-methylbenzene). (d) $n = 2$ (the angle C–O–H is 104°, and the potential for torsional motion about the C–O bond repeats only twice every 360°).

As for an inversion vibration, tunnelling through torsional barriers may occur, resulting in the splitting of energy levels. In the case of an *n*-fold barrier the levels are split into *n* components, but some of these may be degenerate. The example in Figure 6.7(d) shows a three-fold barrier. In principle, tunnelling through the barrier splits the vibrational levels into three components, but two of these are degenerate, having the same energy but different wave functions. For energy levels above a torsional barrier, **free internal rotation** occurs.

If each of the vibrations of a polyatomic molecule is treated in the harmonic oscillator approximation, the vibrational term values, for each vibration *i*, are given by:

$$G(v_i) = \omega_i(v_i + \tfrac{1}{2}) \tag{6.16}$$

This is the same as equation (6.12) for a diatomic molecule, but if vibration *i* is degenerate, $\frac{1}{2}$ is replaced by $d_i/2$, where $d_i$ is the degree of degeneracy of the vibration. For example, $d_i = 2$ for each of the vibrations $\nu_4$ and $\nu_5$ of ethyne in Figure 6.6.

The inclusion of the effects of anharmonicity introduces a further complication. The vibrations are no longer independent. As a result, no expression can be given for the term values for each vibration separately, only an expression for the sum of all vibrational term values, $\Sigma_i G(v_i)$.

When there are no vibrational degeneracies, or unusual vibrational types such as inversion or torsion, this is given by:

$$\Sigma_i G(v_i) = \Sigma_i \omega_i (v_i + \tfrac{1}{2}) + \Sigma_{i \le j} x_{ij} (v_i + \tfrac{1}{2})(v_j + \tfrac{1}{2}) + \ldots \quad (6.17)$$

where vibrations are labelled by $i$ or $j$, and $x_{ij}$ are anharmonic constants. When $i = j$, the term values are similar to those of equation (6.13) for a diatomic molecule. It is the terms involving $x_{ij}$, when $i < j$, which result in the term values for each vibration no longer being independent. For example, for $H_2O$ (see Figure 6.5), contributions to the vibrational term values due to the anharmonic constants $x_{12}$, $x_{23}$ and $x_{13}$ must be taken into account.

## 6.4 Vibration in Excited Electronic States

In Chapter 5 we were concerned not only with the ground electronic state of diatomic and polyatomic molecules, but also with excited electronic states. So far in this chapter, molecular vibration has been assumed to be in the ground electronic state. However, vibration may occur in any electronic state and, using the Born–Oppenheimer approximation, these motions can be considered separately and their associated energies can be added together, as in equation (6.3).

For a diatomic molecule, a typical vibrational potential function associated with an excited electronic state appears qualitatively like that in Figure 6.3 for the ground state. Quantitatively, the dissociation energy, $D_0$, and the equilibrium bond length, $r_e$, are different, the latter very often being greater in the excited state, as shown in the example in Figure 6.9. The fundamental vibration wavenumber, $\omega_e$ of equation (6.13), and the anharmonic constants are also different.

In cases where an excited or a ground electronic state is unstable, as, for example, the $(\sigma_g 1s)^1(\sigma_u 1s)^1$ excited state of $H_2$ or the $(\sigma_g 1s)^2(\sigma_u 1s)^2$ ground state of $He_2$ (see Section 5.1.1), the potential curve appears like that in Figure 6.10. Such a curve does not support any discrete vibrational levels but leads directly to dissociation.

In excited electronic states of polyatomic molecules, there is a potential energy hypersurface, resembling that for the ground state, corresponding to the $3N - 5$ (linear) or $3N - 6$ (non-linear) modes of vibration. Approximate two-dimensional cross-sections can be made in this surface for each vibrational mode. Each of these two-dimensional curves supports vibrational levels.

We shall consider vibration in excited electronic states in more detail in Chapter 11.

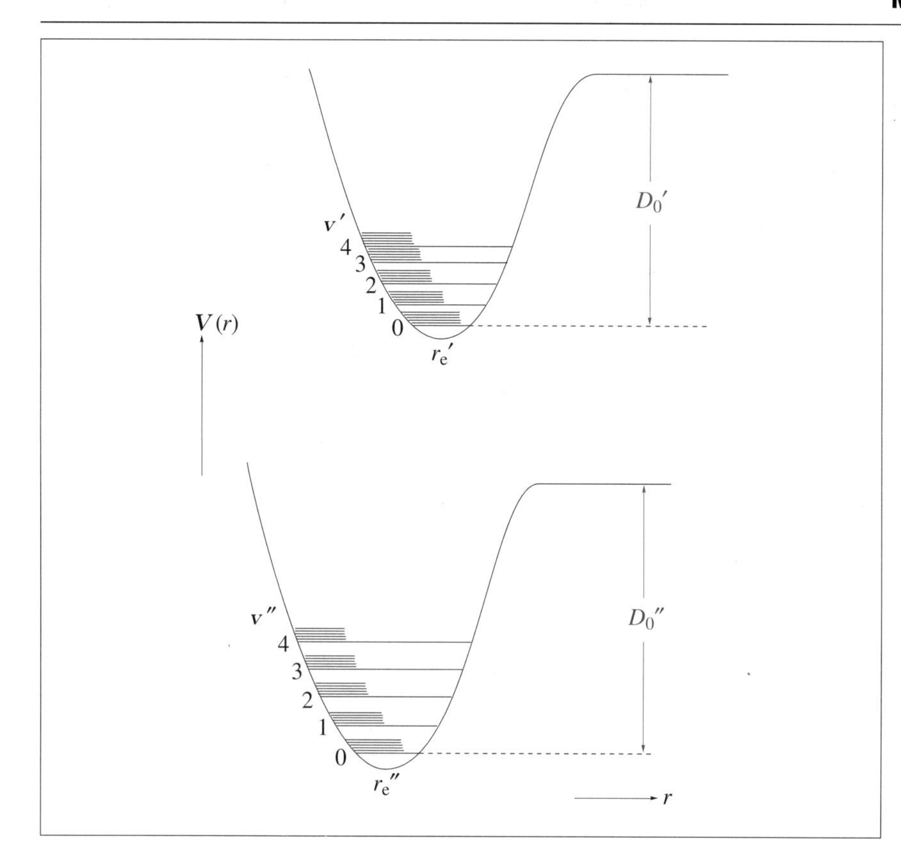


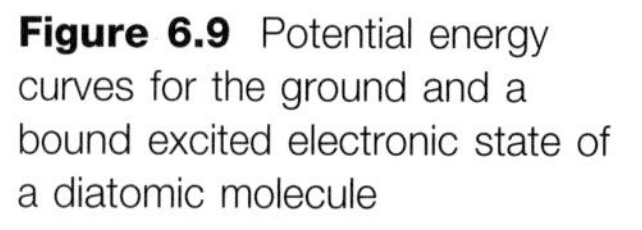

**Figure 6.9** Potential energy curves for the ground and a bound excited electronic state of a diatomic molecule

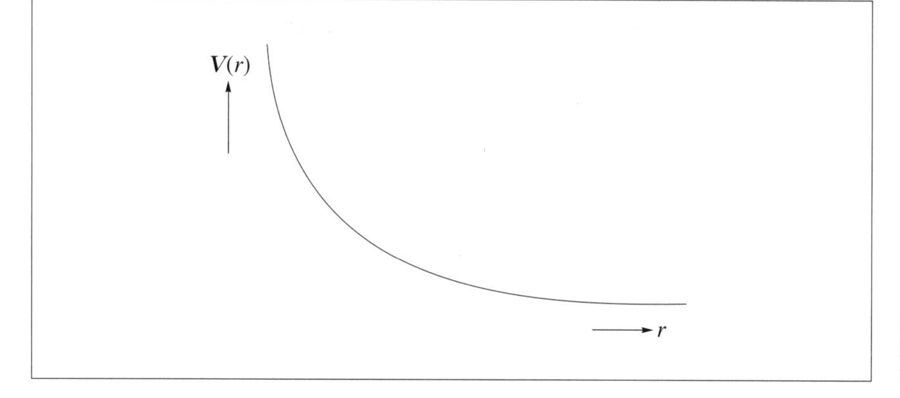


**Figure 6.10** Potential energy curve for an unbound electronic state

## Summary of Key Points

**1.** *Ball and spring model for molecular vibration*

The strength of a spring represents the strength of a bond or the resistance to bending.

**2.** *Harmonic oscillator approximation for a diatomic molecule*
Hooke's law, the concept of a force constant, and vibrational term values. Zero-point energy.

**3.** *Anharmonic oscillator treatment for a diatomic molecule*
Potential curve, dissociation, and vibrational term values. Birge–Sponer extrapolation to give the dissociation energy.

**4.** *Vibrations of polyatomic molecules*
For a linear $N$-atomic molecule, there are $3N - 5$ normal modes of vibration, and $3N - 6$ if it is non-linear. The concept of a group vibration and its approximate transferability from one molecule to another. The concept of a multidimensional potential energy hypersurface to represent all the vibrations. Local mode treatment.

**5.** *Non-dissociative vibrations of polyatomic molecules*
Bending, inversion and torsional vibrations show unusual potential energy curves, none of which leads to dissociation. Those for inversion and torsional vibrations show energy barriers between equivalent conformations.

**6.** *Vibrational term values for polyatomic molecules*
Inclusion of $x_{ij}$ anharmonic constants, when $i < j$, results in the fact that the term values for each vibration cannot be treated independently.

## Problems

**6.1.** Given that the force constant for the $^{1}H^{35}Cl$ molecule is 515.7 N m$^{-1}$, and that $m(^{1}H) = 1.0078$ u and $m(^{35}Cl) = 34.969$ u, calculate, in the harmonic oscillator approximation, the vibration wavenumber of this molecule.

**6.2.** For the $^{12}C^{14}N$ molecule, $\omega_e = 2068.59$ cm$^{-1}$, $x_e = 13.087$ cm$^{-1}$ and $y_e = -0.00909$ cm$^{-1}$. Compare the separations, to four significant figures, of the $v = 1$–0 and 11–10 vibrational energy levels.

**6.3.** Given the following data, calculate the difference, in cm$^{-1}$ and to two decimal places, between the zero-point levels of $^{1}H_2$ and $^{2}H_2$. For $^{1}H_2$, $\omega_e = 4401.21$ cm$^{-1}$, $x_e = 121.34$ cm$^{-1}$ and $y_e = 0.813$ cm$^{-1}$; for $^{2}H_2$, $\omega_e = 3115.50$ cm$^{-1}$, $x_e = 61.82$ cm$^{-1}$ and $y_e = 0.562$ cm$^{-1}$. Comment on the significance of this difference for the rates of reactions involving the breaking of an X–$^{1}$H or X–$^{2}$H bond.

# 7
# Molecular Rotation

**Aims**

In this chapter you will be introduced to

- Rotational energy levels (term values) for diatomic and linear polyatomic molecules treated in the rigid rotor approximation
- Effects of centrifugal distortion on these energy levels
- The principal moments of inertia of a molecule
- Definitions of symmetric, spherical and asymmetric top molecules
- Rotational energy levels of symmetric, spherical and asymmetric top molecules

## 7.1 Introduction

We have seen in Chapters 5 and 6 that molecules have internal energy due to the orbital motions of the electrons and the vibrational motions of the nuclei. Both types of energy can take only discrete values, *i.e.* they are quantized. A further source of energy is in the overall rotation of the molecule. This energy is quantized also, but quanta of rotational energy are, in general, small compared to those of vibrational energy which are, in turn, small compared to those of electronic energy.

When considering their rotational energy, it is useful to divide molecules into five categories: diatomic, linear, symmetric tops, spherical tops and asymmetric tops.

## 7.2 Diatomic and Linear Polyatomic Molecules

### 7.2.1 Rigid Rotor Approximation

Equation (6.3) shows that, to an approximation which is generally a good one, electronic, vibrational and rotational energies, $E_e$, $E_v$ and $E_r$, can be treated independently. To obtain $E_r$ it is necessary to solve the Schrödinger equation of equation (3.19), but this is well beyond the scope of this book; we shall require only the results.

In considering molecular rotation, it is a useful approximation to regard the bonds as rigid rods, and the molecule as a **rigid rotor**. Then, the rotational energy for a diatomic or linear polyatomic molecule is given by:

$$E_r = (h^2/8\pi^2 I)J(J + 1) \quad (7.1)$$

where the rotational quantum number $J = 0, 1, 2, 3, \ldots$ . The quantity $I$ is the **moment of inertia** of the molecule for rotation about any axis through the **centre of mass** and at 90° to the internuclear axis. This is illustrated, for a homonuclear and a heteronuclear diatomic molecule, in Figure 7.1(a) and (b), respectively. The moment of inertia is given by:

$$I = \mu r^2 \quad (7.2)$$

in which $\mu$ is the reduced mass of the two nuclei of masses $m_1$ and $m_2$:

$$\mu = m_1 m_2/(m_1 + m_2) \quad (7.3)$$

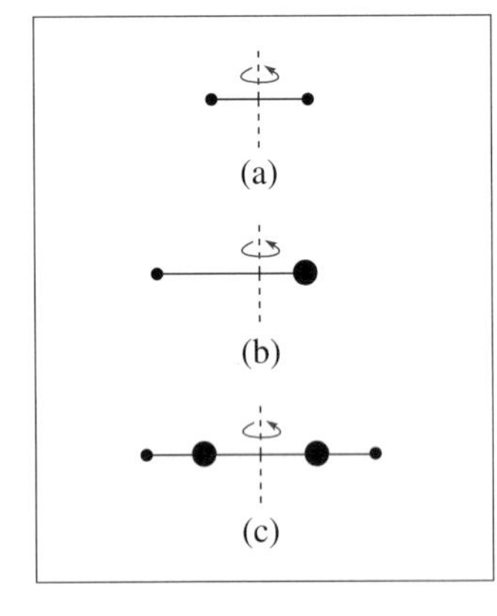


**Figure 7.1** Rotation of (a) a homonuclear diatomic, (b) a heteronuclear diatomic and (c) a symmetrical linear polyatomic molecule

and $r$ is the bond length.

For a linear polyatomic molecule also, rotation can occur about any axis through the centre of mass and at 90° to the internuclear axis. In the case of a symmetrical, linear polyatomic molecule, such as ethyne (acetylene), H–C≡C–H (Figure 7.1c), the centre of mass is at the centre of the molecule. In an unsymmetrical molecule, such as hydrogen cyanide, H–C≡N, this is not so. In both types of molecule, the moment of inertia is given by:

$$I = \Sigma m_i r_i^2 \quad (7.4)$$

*i.e.* the sum of all quantities $m_i r_i^2$, where $m_i$ is the mass of atom $i$ and $r_i$ its distance from the centre of mass.

Figure 7.2 shows a typical set of rotational energy levels, given by equation (7.1), for a diatomic or linear polyatomic molecule, diverging smoothly as the quantum number $J$ increases.

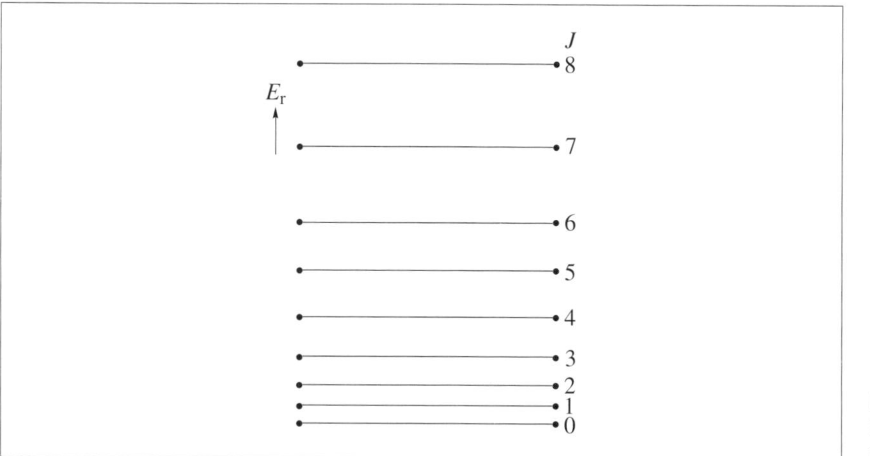


**Figure 7.2** Rotational energy levels for a diatomic or linear polyatomic molecule

In any kind of spectroscopy which involves rotational energy levels it is usually the wavenumbers or frequencies, rather than energies, that are measured. The energy levels of equation (7.1) can be converted to frequencies (see equation 2.9), giving:

$$F(J) = E_r/h = (h/8\pi^2 I)J(J + 1) = BJ(J + 1) \tag{7.5}$$

or to wavenumbers, giving:

$$F(J) = E_r/hc = (h/8\pi^2 cI)J(J + 1) = BJ(J + 1) \tag{7.6}$$

The quantities $F(J)$ are **rotational term values** and $B$ is the **rotational constant**. It is unfortunate, but traditional, that the symbol $B$ is used whether it refers to a rotational constant in terms of wavenumber or frequency. Because $B$ is proportional to $1/\mu$, the term values increase more slowly, as $J$ increases, for heavier molecules.

The unit of Å, the **ångström**, is not an SI unit but is frequently encountered as a measure of bond length. The preferred SI unit for bond lengths is either the nanometre (nm), where 1 Å = 0.1 nm, or, rather less so, the picometre (pm), where 1 Å = 100 pm.

### Worked Problem 7.1

**Q** Calculate, to four significant figures and with units of $cm^{-1}$, the rotational constant $B$ for the molecules (a) $^{12}C^1H$ and (b) $^{127}I_2$, given that $m(^{12}C) = 12.000$ u, $m(^1H) = 1.0078$ u, $r(CH) = 1.1199$ Å, $m(^{127}I) = 126.90$ u and $r(I_2) = 2.6663$ Å, where 1 Å = $10^{-10}$ m. Convert these $B$-values to units of GHz (1 GHz = $10^9$ Hz).

**A** (a) The reduced mass is given by:

$$\begin{aligned}\mu &= m(^{12}C)m(^1H)/[m(^{12}C) + m(^1H)]\\ &= 12.000 \times 1.0078/(12.000 + 1.0078)\ \text{u}\\ &= 0.92972\ \text{u} = 0.92972 \times 1.6605 \times 10^{-27}\ \text{kg}\\ &= 1.5438 \times 10^{-27}\ \text{kg}\end{aligned}$$

Then, from equation (7.2):

$$I = \mu r^2 = 1.5438 \times 10^{-27} \times (1.1199 \times 10^{-10})^2 \text{ kg m}^2$$
$$= 1.9362 \times 10^{-47} \text{ kg m}^2$$

Therefore, from equation (7.6):

$$B = h/8\pi^2 cI$$
$$= 6.6261 \times 10^{-34} \text{J s}/8\pi^2 \times 2.9979 \times 10^{10} \text{ cm s}^{-1} \times 1.9362 \times 10^{-47} \text{ kg m}^2$$
$$= 14.46 \text{ cm}^{-1}$$

Multiply by $c$, the speed of light, to give $B$ with units of frequency:
$\therefore\ B = 14.458 \times 2.9979 \times 10^{10}$ cm s$^{-1}$ = $4.334 \times 10^{11}$ s$^{-1}$ = 433.4 GHz
(b) Following the same method as in (a) gives:
$\mu = m(^{127}\text{I})/2 = 1.0536 \times 10^{-25}$ kg; $I = 7.4902 \times 10^{-45}$ kg m$^2$; $B =$ 0.03737 cm$^{-1}$ or 1.120 GHz

Note how much smaller is the value of $B$ for the heavier $I_2$ molecule.

### 7.2.2 Treatment as a Non-rigid Rotor

In Chapter 6 we saw that the bonds between atoms should be thought of as springs rather than rigid rods. This applies when considering rotation also. The rigid rotor approximation in Section 7.2.1 applies only to low speeds of rotation, *i.e.* to energy levels with low values of $J$. At higher speeds of rotation the spring-like bonds expand due to the centrifugal forces which throw the atoms outwards. This slight expansion of the bond in a diatomic molecule (or bonds in a linear polyatomic molecule) is known as **centrifugal distortion**. It is taken account of by adding higher order terms in $J(J + 1)$ to the term values of equation (7.5) or (7.6) to give:

$$F(J) = BJ(J + 1) - DJ^2(J + 1)^2 + \ldots \tag{7.7}$$

where $D$ is the **centrifugal distortion constant**. Equation (7.7) indicates that further terms may be included, but these are too small to concern us here.

**Worked Problem 7.2**

**Q** For the $^1\text{H}^{79}\text{Br}$ molecule, $B$ = 250.36 GHz and $D$ = 10.44 MHz. Compare the $J$ = 1–0 and 11–10 term value separations, treating the molecule as (a) a rigid rotor and (b) a non-rigid rotor (note that 1 GHz = $10^3$ MHz.)

**A** Using (a) equation (7.5) and (b) equation (7.7) gives:

| | $F(J)$/GHz | |
|---|---|---|
| $J$ | (a) | (b) |
| 0 | 0 | 0 |
| 1 | 500.72 | 500.68 |
| 10 | 27 540 | 27 414 |
| 11 | 33 048 | 32 866 |

The $J$ = 1–0 separation is reduced by only 0.04 GHz by inclusion of the non-rigid rotor term, whereas the $J$ = 11–10 separation is reduced from 5508 to 5452 GHz. This result shows the increasing importance of centrifugal distortion at high values of $J$.

## 7.3 Non-linear Polyatomic Molecules

### 7.3.1 Introduction

In order to consider the rotational energy levels of non-linear polyatomic molecules, we must classify the molecules according to their **principal moments of inertia**. In general, the moment of inertia $I$ (with reference to rotation of the molecule about *any* axis passing through the centre of mass) is given by equation (7.4), where $r_i$ is the distance of an atom of mass $m_i$ from the axis. The summation, $\Sigma_i$, is over all the atoms. In any non-linear molecule there are two unique axes about which the moment of inertia is a minimum or a maximum; conventionally, these are labelled the *a*- or *c*-axis, respectively. These are two of the so-called **principal axes**, and are at 90° to each other. The corresponding moments of inertia, $I_a$ and $I_c$, are two of the principal moments of inertia. There is a third principal moment of inertia, $I_b$, about the *b*-axis, which is at 90° to the *a*- and *c*-axes. All three axes pass through the centre of mass of the molecule. In general:

$$I_a \leq I_b \leq I_c \tag{7.8}$$

### 7.3.2 Symmetric Top Molecules

A **symmetric top** (or **symmetric rotor**) molecule is defined as one which has two equal principal moments of inertia. From equation (7.8) we can see that there are two possibilities: (a) that $I_a < I_b = I_c$ and (b) that $I_a =$

$I_b < I_c$. Such molecules are called (a) **prolate** and (b) **oblate** symmetric tops.

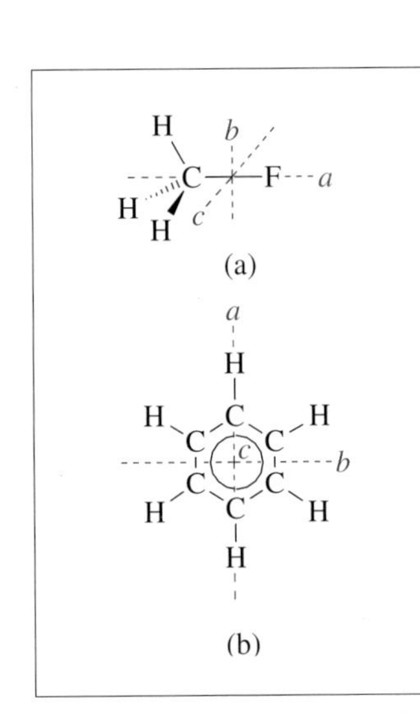


**Figure 7.3** (a) Fluoromethane, a prolate symmetric top, and (b) benzene, an oblate symmetric top

Figure 7.3(a) shows an example of a prolate and Figure 7.3(b) of an oblate symmetric top. In the fluoromethane molecule, in Figure 7.3(a), it is clear that the moment of inertia about the *a*-axis is the minimum moment of inertia. This is because the only atoms which are not on this axis, and which therefore contribute to the moment of inertia about this axis, are the light hydrogen atoms. It would require a little trigonometry to show that $I_b = I_c$.

The benzene molecule, shown in Figure 7.3(b), is an oblate symmetric top, with the axis of maximum moment of inertia, the *c*-axis, perpendicular to the figure. Again, it would require trigonometry to show that $I_a = I_b$.

For non-linear molecules there are now three rotational constants. Analogous to the *B* rotational constant for a diatomic molecule (see equations 7.5 and 7.6), these rotational constants are given by:

$$A = h/8\pi^2 I_a; \quad B = h/8\pi^2 I_b; \quad C = h/8\pi^2 I_c \tag{7.9}$$

with units of frequency, or by:

$$A = h/8\pi^2 cI_a; \quad B = h/8\pi^2 cI_b; \quad C = h/8\pi^2 cI_c \tag{7.10}$$

with units of wavenumber.

In symmetric tops a second quantum number, *K*, is introduced in addition to *J*. The equation for the rotational term values, $F(J,K)$, analogous to that of equation (7.5) (or 7.6) for a diatomic molecule, becomes:

$$F(J,K) = BJ(J+1) + (A - B)K^2 \tag{7.11}$$

for a prolate symmetric top, for which $C = B$, and:

$$F(J,K) = BJ(J+1) + (C - B)K^2 \tag{7.12}$$

for an oblate symmetric top, for which $A = B$. The quantum number *K* can only be less than or equal to *J*, *i.e.* $K = 0, 1, 2, 3, \ldots J$.

Figure 7.5(a) and (b) show sets of rotational energy levels for a prolate and an oblate symmetric top, respectively. For each value of *K* there is a stack of *J*-dependent levels which resemble those for a diatomic molecule, except that the lowest level in each stack is that with $J = K$. The main difference between the levels for the two types of symmetric top is that, for a particular value of *J*, the levels diverge for a prolate symmetric top but converge for an oblate symmetric top. The reason for this is apparent from equations (7.11) and (7.12). For a prolate symmetric

## Box 7.1 Vector Representation of Rotational Angular Momentum

Rotational angular momentum is a vector quantity, having magnitude and direction. In a diatomic or linear polyatomic molecule, rotation is about any axis through the centre of mass and perpendicular to the internuclear axis. Figure 7.4(a) shows that the vector, $\boldsymbol{P}$, lies along that axis. The length of the vector is $J(J + 1)\hbar$.

In general, a prolate or oblate symmetric top may rotate about *any* axis through the centre of mass, and the vector, $\boldsymbol{P}$, lies along that axis, as shown in Figure 7.4(b) for a prolate symmetric top. The magnitude of the vector is $J(J + 1)\hbar$. Figure 7.4(b) also shows that the component of the vector along the unique *a*-axis (the *c*-axis in an oblate symmetric top) is $\boldsymbol{P}_a$ ($\boldsymbol{P}_c$ in an oblate symmetric top). The magnitude of this component is $K\hbar$. It is clear from the figure that the quantum number $K \leq J$.

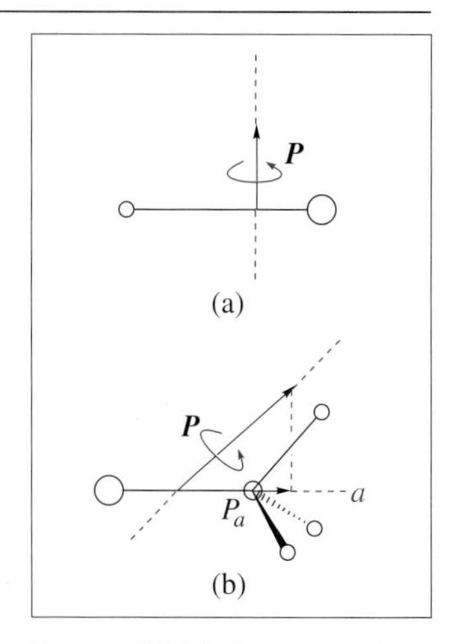


**Figure 7.4** Vector representation of rotation in (a) a heteronuclear diatomic molecule and (b) a prolate symmetric top molecule, such as fluoromethane

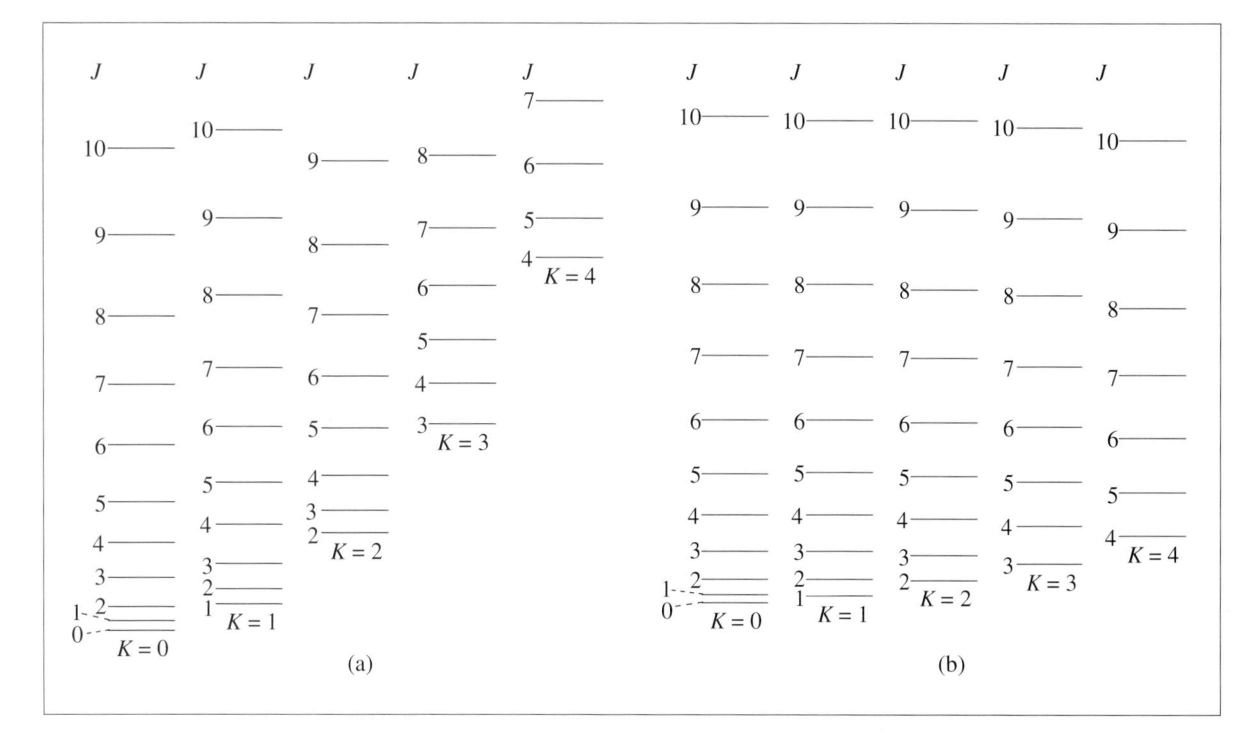


**Figure 7.5** Rotational energy levels for (a) a prolate and (b) an oblate symmetric top molecule

top the value of $A - B$ is positive, but for an oblate symmetric top the value of $C - B$ is negative.

As with a diatomic or linear polyatomic molecule, centrifugal distortion occurs increasingly as the rotational quantum number(s) increase.

To take account of these relatively small effects, extra terms must be added to the rotational term value expressions of equations (7.11) and (7.12), but they will not concern us here.

### 7.3.3 Spherical Top Molecules

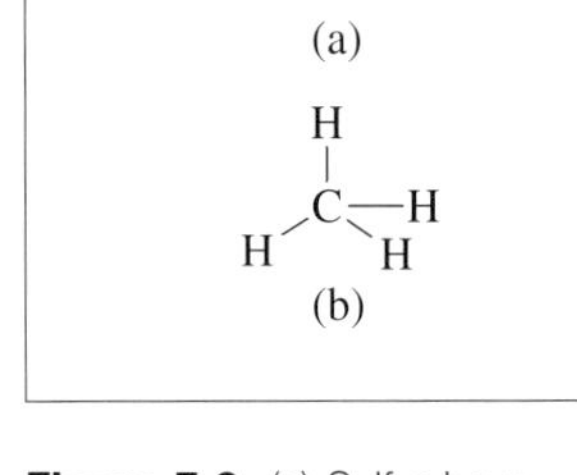


**Figure 7.6** (a) Sulfur hexafluoride and (b) methane are spherical top molecules

For any molecule having the structure of a regular octahedron, such as $SF_6$ (Figure 7.6a), or a regular tetrahedron, such as $CH_4$ (Figure 7.6b), all three principal moments of inertia are equal, *i.e.* $I_a = I_b = I_c$. Such molecules are **spherical tops** (or **spherical rotors**). The rotational term values are given by:

$$F(J) = BJ(J + 1) \tag{7.13}$$

when the effects of centrifugal distortion are neglected. The rotational constant, $B$, is given by the same expression as in equation (7.9) or (7.10). The term value expression is the same as that for a diatomic or linear polyatomic molecule, and the pattern of rotational energy levels is the same as that in Figure 7.2.

### 7.3.4 Asymmetric Top Molecules

For an **asymmetric top** (or **asymmetric rotor**) molecule, all the principal moments of inertia are unequal:

$$I_a \neq I_b \neq I_c \tag{7.14}$$

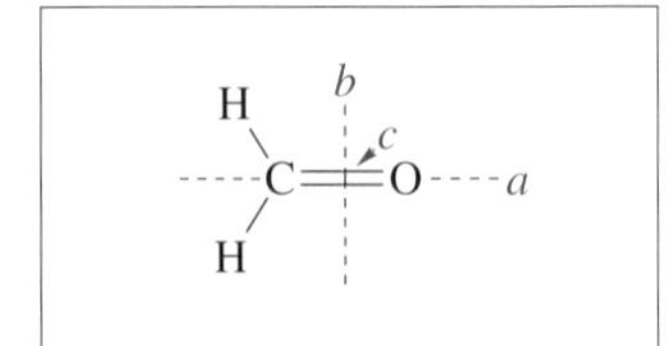


**Figure 7.7** Methanal (formaldehyde) is an asymmetric top molecule

The vast majority of molecules are asymmetric tops, of which methanal (formaldehyde), shown in Figure 7.7, is an example.

Unfortunately, there are no simple formulae for the rotational term values for asymmetric tops. $J$ is still a good quantum number, *i.e.* takes integral values, but $K$ is not. Approximate expressions for those with low values of $J$ have been derived, and also for molecules which approximate to either a prolate or oblate symmetric top, *i.e.*:

$$I_a > I_b \approx I_c \quad \text{or} \quad I_a \approx I_b < I_c \tag{7.15}$$

Accurate term values can be obtained only by the use of matrix algebra, which is beyond the scope of this book.

## Summary of Key Points

1. *Rotational term values for diatomic and linear polyatomic molecules*
What is meant by a rigid rotor. Definition of moment of inertia. Rotational term value expressions for diatomic and linear polyatomic molecules treated as rigid rotors.

2. *Centrifugal distortion*
Why it occurs, and its effect on the rotational term values.

3. *Principal axes and principal moments of inertia of a non-linear molecule*
How these lead to the definition of various classes of non-linear molecules.

4. *Symmetric tops (or rotors)*
Term value expressions, neglecting centrifugal distortion.

5. *Spherical tops (or rotors)*
Term value expression, neglecting centrifugal distortion.

6. *Asymmetric tops (or rotors)*
There are no simple, accurate expressions for the term values, even in the rigid rotor approximation.

## Problems

**7.1.** Calculate the principal moment of inertia of the regular octahedral molecule $^{238}U^{19}F_6$, given that $r$(U–F) = 1.9993 Å, where 1 Å = $10^{-10}$ m, and $m(^{19}F)$ = 18.998 u. Hence, calculate the $B$-value in units of $cm^{-1}$. How would this value be affected for other isotopes of uranium?

**7.2.** Trifluoromethyl iodide, $CF_3I$, is a prolate symmetric top, with rotational constants $A$ = 0.1910 $cm^{-1}$ and $B$ = 0.05081 $cm^{-1}$. Neglecting centrifugal distortion, calculate all the rotational energy levels for $J$ = 2. Make a similar calculation for the oblate symmetric top ammonia, $NH_3$, for which $B$ = 9.443 $cm^{-1}$ and $C$ = 6.196 $cm^{-1}$, and make a qualitative comparison with those for $CF_3I$.

# 8
# How Spectra are Obtained

**Aims**

In this chapter the reader will be introduced, in outline, to the experimental techniques used in rotational, vibrational, electronic and Raman spectroscopy. By the end of this chapter you should:

- Be aware of the features common to most experimental methods for obtaining absorption spectra
- Be familiar with the definition of absorbance, and the Beer–Lambert law
- Be aware of the sources of radiation, methods of dispersion, materials for cell windows and detectors used in microwave, infrared, visible and ultraviolet spectroscopy
- Be familiar with Raman scattering and the experimental methods used for its observation
- Know what effects limit spectral line widths
- Be aware of the various phases in which spectra are observed

## 8.1 Microwave, Infrared, Visible and Ultraviolet Spectroscopy

In this book we are concerned primarily with the patterns of transitions in various types of spectra and the information which may be obtained from them. Most of the experimental techniques employed to obtain these spectra will be described in outline only, except for Raman spectroscopy for which the technique is very different from those used for microwave, infrared, visible and ultraviolet spectroscopy.

The majority of spectra to be discussed will be absorption spectra. The general experimental arrangement to obtain such a spectrum is shown, symbolically, in Figure 8.1. The radiation from the source passes

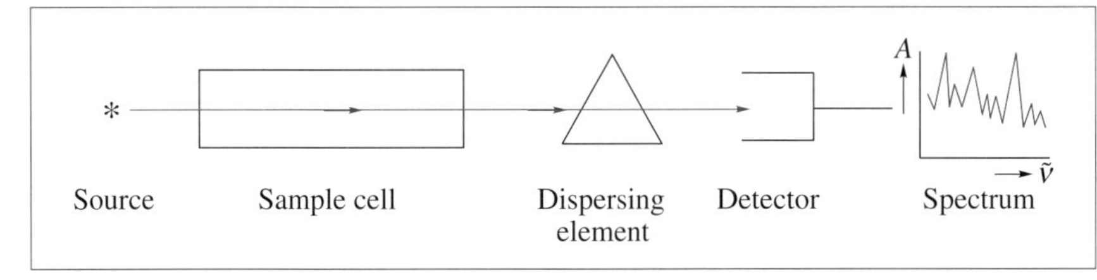


**Figure 8.1** Symbolic representation of the principal components used in obtaining an absorption spectrum

through the sample cell, containing the absorbing sample, to a **dispersing element**. This element disperses and resolves the radiation falling on it. It may be a prism, as shown, but is usually a diffraction grating; these dispersing elements have been described in Sections 1.1–1.2. The radiation is detected as the wavelength or wavenumber (see equation 2.4) is changed smoothly by rotating the prism or diffraction grating. The spectrum is shown as a plot of **absorbance**, $A$, against wavenumber, $\tilde{\nu}$. Absorbance is of the greatest importance in electronic (usually visible or ultraviolet) spectroscopy and is a measure of the degree of absorption by the sample. It is defined as:

$$A = \log_{10}(I_0/I) \tag{8.1}$$

where, as shown in Figure 8.2, $I_0$ is the intensity of radiation of a particular wavelength entering the cell of length $l$, and $I$ is the intensity of the radiation after absorption has taken place.

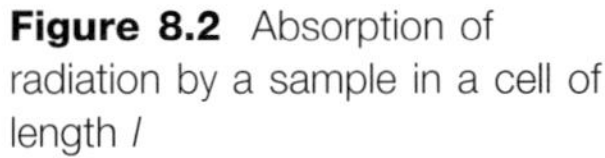

**Figure 8.2** Absorption of radiation by a sample in a cell of length $l$

According to the **Beer–Lambert law** the absorbance is proportional to $l$ and to the concentration, $c$, of the sample in the cell:

$$A = \varepsilon(\tilde{\nu})cl \tag{8.2}$$

The quantity $\varepsilon$ is a measure of the absorbing power of the sample at a particular wavenumber, $\tilde{\nu}$, and is called the **molar absorption coefficient** or the **molar absorptivity**. Since absorbance has no units, the units of $\varepsilon$ must be those of concentration$^{-1}$ × length$^{-1}$, and are usually mol$^{-1}$ dm$^3$ cm$^{-1}$ (it is usual for the cell length to be measured in centimetres).

In many older texts, the molar absorption coefficient is referred to as the **molar extinction coefficient** or, simply, as the extinction coefficient.

### Worked Problem 8.1

**Q** At a wavenumber of 33 400 $cm^{-1}$ a sample, contained in a cell $6.45 \times 10^{-3}$ m long, absorbs 76.4% of the incident radiation. The sample is in the liquid phase in solution with a concentration of $1.76 \times 10^{-4}$ mol $dm^{-3}$. What is the value of the molar absorption coefficient, with units of $mol^{-1}$ $dm^3$ $cm^{-1}$?

**A** $I_0/I = 100/(100 - 76.4) = 4.237$
$\therefore A = \log_{10}(I_0/I) = 0.627$
$\therefore \varepsilon = 0.627/1.76 \times 10^{-4}$ mol $dm^{-3} \times 0.645$ cm
$= 5520$ $mol^{-1}$ $dm^3$ $cm^{-1}$

The absorption intensity of the band shown in Figure 8.3 is the total area under the curve. However, the value of the molar absorption coefficient, $\varepsilon_{max}$, at the maximum absorbance is often used as an approximate measure of the total absorption intensity. The quantity $\varepsilon_{max}$ is very useful when spectroscopy is used for analytical purposes, usually in the liquid phase in solution. If $\varepsilon_{max}$ is known for the spectrum of a molecule in a particular solvent, and the cell length $l$ is known, the concentration follows simply from the measured absorbance $A$ and equation (8.2).

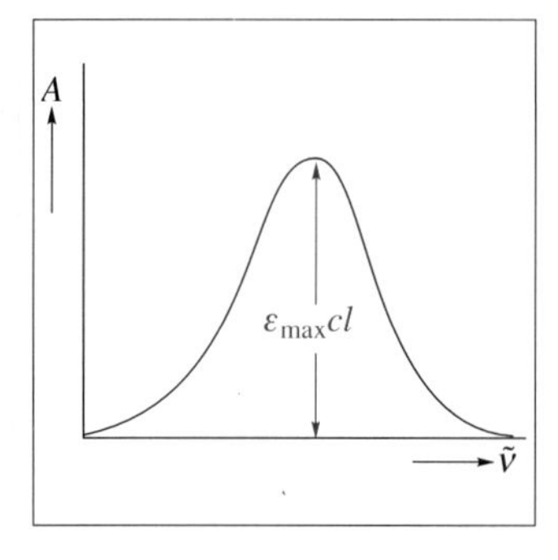


**Figure 8.3** A broad absorption band

Figure 8.4 shows the visible/ultraviolet absorption spectrum of *trans*-dimethyldiimide [(*E*)-1,2-dimethyldiazene] in solution in hexane. Typical of many such spectra in solution, it shows very little structure. Nevertheless, measurement of the absorbance exhibited by either, or both, of the maxima at 42 700 and 28 200 $cm^{-1}$ can be used to measure the concentration of the *trans*-dimethyldiimide, provided that $\varepsilon_{max}$ is known for these maxima.

### Worked Problem 8.2

**Q** The absorbances $A_1$ and $A_2$, for the maxima at 42 700 and 28 200 $cm^{-1}$ in the spectrum in Figure 8.4, are 2.55 and 0.55, respectively. Estimate the corresponding values of $\varepsilon_{max}$, using the scale carefully. Given that the length of the absorption cell was 5.0 cm, use both maxima to calculate the concentration of the solute.

**A** The estimated values of $\varepsilon_{max}$, from the graph, for the maxima at 42 700 and 28 200 $cm^{-1}$ are 60 and 13, respectively. Then, using equation (8.2):
$c = A_1/60$ $mol^{-1}$ $dm^3$ $cm^{-1} \times 5.0$ cm $= 8.5 \times 10^{-3}$ mol $dm^{-3}$
and $c = A_2/13$ $mol^{-1}$ $dm^3$ $cm^{-1} \times 5.0$ cm $= 8.5 \times 10^{-3}$ mol $dm^{-3}$

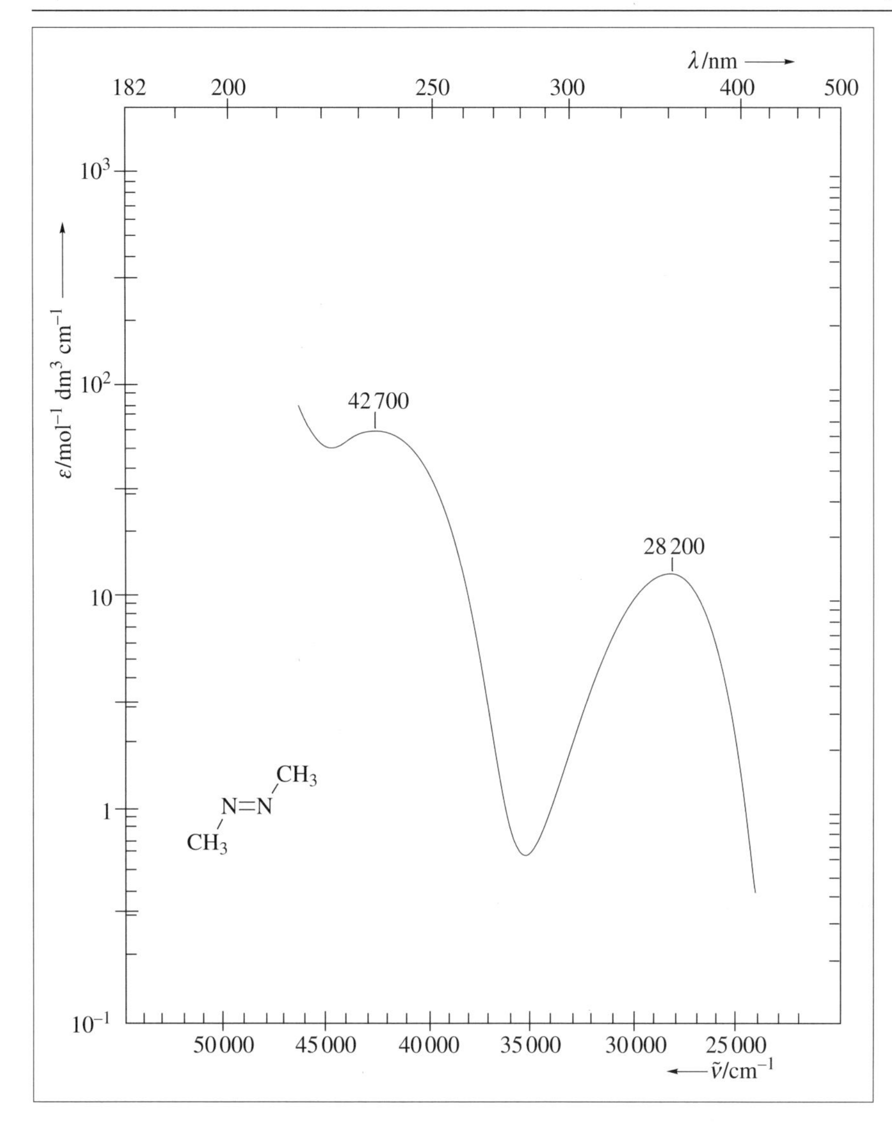


Note that the vertical scale is logarithmic.

**Figure 8.4** Visible and near-ultraviolet absorption spectrum of *trans*-dimethyldiimide in solution in hexane (after G. Kortüm and H. Rau, *Ber. Bunsenges. Phys. Chem.*, 1964, **68**, 973)

Quantitative absorption intensities are much less important in microwave and infrared spectroscopy, and will not be discussed here.

The experimental methods involved in obtaining microwave, infrared, visible or ultraviolet absorption spectra each have unique features, but are generally based on the arrangement in Figure 8.1. The reader is referred to Section 2.1 for definitions of the various regions of the electromagnetic spectrum.

The source of radiation used for microwave and millimetre wave spectroscopy is a **klystron** or a **backward wave oscillator**. These are electronic devices that produce radiation which is highly monochromatic, *i.e.* having a very narrow band width, and are smoothly tuneable through a specific wavelength range. The radiation differs from infrared, visible and ultraviolet radiation in that it travels inside metal tubing. This tubing has a rectangular cross-section. It is known as **waveguide**, and guides the radiation, around corners if necessary, from the source to the detector. The absorption cell forms a part of the waveguide, having **mica** windows at each end to contain the sample. Because the source is monochromatic and tuneable, a dispersing element is not required. The type of detector used is a **crystal diode rectifier**, an electronic device.

Far-infrared spectroscopy is the most difficult of the infrared spectroscopies, mainly due to the fact that water vapour, which is always present in air, absorbs the radiation very strongly. Water vapour must be excluded from the complete optical path from the source, such as a mercury discharge in a quartz envelope, to the detector. This may be partly achieved by flushing the whole spectrometer with dry nitrogen, but complete evacuation is preferable. Polymeric materials, such as poly(phenylethene) (polystyrene), transmit far-infrared radiation and are used for cell windows. One type of detector which is used is a **Golay cell**, which relies on the heating of an aluminium film inside a flexible container filled with xenon gas. A light beam falls on a mirror deposited on the outside of the container and is deflected according to the degree of heating in the cell, *i.e.* according to the intensity of the far-infrared radiation falling on it. Other types of detectors include thermocouples and semiconductors.

The dispersing element which is used for far-infrared radiation is either a diffraction grating or an **interferometer**. The details of how an interferometer works will not concern us here but, briefly, the beam of radiation emerging from the absorption cell is split into two parts. One part strikes a fixed mirror in the interferometer and the other part strikes a moving mirror. When the two parts are recombined, interference occurs between them. The degree of interference depends on the difference between the lengths of the paths traversed by the two beams. Higher resolution (see Section 1.2) may be achieved with an interferometer than with a diffraction grating. Another advantage of using an interferometer is that it allows the observation of all the spectrum all the time, the so-called **Fellgett advantage**, whereas the use of a diffraction grating rejects most of the spectrum most of the time. For this reason, an interferometer is used in preference to a diffraction grating when the speed of recording a spectrum is important, or for the observation of weak spectra.

In mid- and near-infrared spectroscopy, evacuation of the optical line is unnecessary. The source is either a heated **Nernst filament**, consisting of mixed rare earth oxides, or a heated, silicon carbide **Globar**. The dispersing element may be either a diffraction grating or an interferometer. Cell windows are made from either sodium chloride or potassium bromide. If the sample is in solution, the solvent must not absorb the radiation in the region of interest. Solvents such as carbon disulfide and tetrachloroethene are used. A solid sample may be ground into a fine powder and made into a mull with **Nujol** (a liquid paraffin). Alternatively, it may be ground together with potassium bromide and compressed to form a **KBr disk**. Types of detectors used, such as a Golay cell, thermocouple or semiconductor, are the same as those for the far infrared.

For visible or near-ultraviolet spectroscopy the source may be a **tungsten filament lamp** or a **deuterium discharge lamp**, respectively. A more intense source is a high-pressure **xenon arc lamp**. Glass (for the visible region only) or fused quartz are used for cell windows. For samples in solution, the solvent must be transparent in the region of interest; a saturated hydrocarbon, such as pentane or hexane, is suitable.

In addition to observing spectra by an absorption process, they may also be observed by an emission process (see Section 1.3). Such emission spectra are observed most commonly in the visible and near-ultraviolet regions. Preceding any emission process, energy must be transferred to the atom or molecule to get it into an excited electronic state. This energy transfer may be indiscriminate, as in the case of an electrical discharge induced in the gaseous material at very low pressure. One example of this is the sodium discharge lamp (see Section 1.3), in which many different wavelengths in the emission spectrum are observed. An electrical discharge in very low-pressure nitrogen gas produces a complex emission spectrum of $N_2$, consisting of discrete bands throughout the visible and near-ultraviolet regions.

Alternatively, a monochromatic (single wavelength) source of energy may be used to excite the atom or molecule into a single excited energy level, from which an emission spectrum may be observed. Single intense lines from an atomic discharge, such as is produced from a mercury discharge lamp, have been used, but a **laser** is a much more useful source. Not only is the radiation from a laser monochromatic, it also contains a much narrower band of wavelengths than is contained in an atomic line. In addition, the radiation is very much more intense, and there is a more flexible choice of wavelengths.

In the far-ultraviolet region, the principal experimental problem follows from the fact that oxygen in the air absorbs strongly at wavelengths below 185 nm, necessitating evacuation of the complete optical line from the source to the detector. Sources of far-ultraviolet radiation include microwave-induced discharges in argon, krypton or xenon, which cover the range 200–105 nm. The only window material that can be used for the absorption cell is lithium fluoride, which absorbs radiation below 105 nm. At these wavelengths, differential pumping must be used, instead of cell windows, to prevent the sample leaking from the absorption cell into the rest of the optical line. Detection of the radiation is usually by a photomultiplier.

In **differential pumping**, the sample is fed continuously into the sample "cell". In order to maintain a very low pressure on either side of the "cell", the sample is rapidly pumped away at the two positions where cell windows would normally be.

## 8.2 Raman Spectroscopy

As was mentioned briefly in Section 2.5, Raman spectroscopy involves neither absorption nor emission, but scattering of radiation by the sample.

When radiation falls on a gaseous sample it may be absorbed, but if the wavelength of the radiation does not correspond to that of an absorption process, it is scattered. The scattered radiation is mostly of unchanged wavelength, and this is known as **Rayleigh scattering**. The intensity, $I_s$, of this scattering varies strongly with wavelength:

$$I_s \propto \lambda^{-4} \tag{8.3}$$

One result of this preferential scattering at shorter wavelengths is that a cloudless sky appears blue. The radiation observed is scattered radiation from particles in the atmosphere, and equation (8.3) shows that the intensity is greatest at the shortest wavelengths.

A very small proportion of scattered light is of slightly increased or decreased wavelength, and is called **Raman scattering**. That of increased or decreased wavelength is known as **Stokes** or **anti-Stokes Raman scattering**, respectively. It is the observation of this weak Raman scattering that forms the basis of Raman spectroscopy.

Lasers are ideal sources of radiation for obtaining Raman spectra: they are monochromatic, which allows the Raman scattering to be separated from the Rayleigh scattering, and are very intense. A helium–neon or an argon ion laser, in which emission is from Ne or $Ar^+$ at wavelengths of 632.8 or 514.5 nm, respectively, is often used. The dispersing element is usually a diffraction grating, but the use of an interferometer (see Section 8.1) has enabled the use of a neodymium–YAG laser source in which emission is from $Nd^{3+}$ at a wavelength of 1064 nm, in the near infrared.

In a **neodymium–YAG laser** the neodymium ions are embedded in a matrix of yttrium aluminium garnet ($Y_3Al_5O_{12}$), for which the acronym is YAG.

The wavelength dependence of the intensity of scattering intensity, illustrated by equation (8.3), would seem to mitigate against the use of a wavelength as high as 1064 nm. However, there are compensating advantages. The use of an interferometer allows the observation of all the spectrum all the time, and is ideal for collecting weak spectra. A further advantage of using a wavelength of 1064 nm is that it is much less likely to fall in a region in which the sample absorbs; absorption of the source radiation, and subsequent emission, precludes the observation of the very weak Raman scattering.

There are several possible optical arrangements of the sample cell in order to optimize the production and collection of the Raman scattering. Figure 8.5 shows one which is commonly used. The lens $L_1$ focuses the laser beam into the cell, C. The concave mirror, $M_1$, sends the emerging beam back into the cell to produce further scattering. The scattered

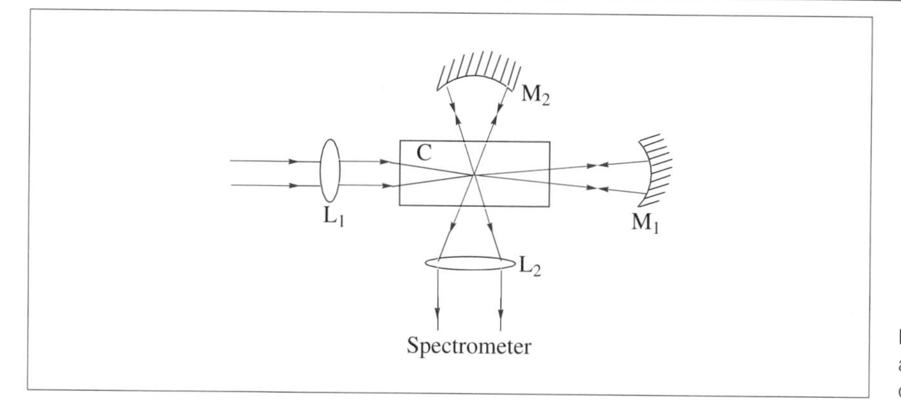


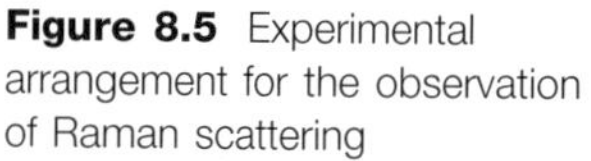
**Figure 8.5** Experimental arrangement for the observation of Raman scattering

radiation is directed by the lens $L_2$ into the dispersing device of the spectrometer; the mirror $M_2$ serves to double the intensity of the scattered radiation which is collected.

The detector used is one which is suitable for the wavelength range in which the laser operates.

## 8.3 Spectral Line Widths

The width of a sharp spectral line, such as the D lines of sodium illustrated in Figure 1.3, may be limited by the width of the entrance slit of the spectrometer, as discussed in Section 1.2. This limitation can be removed by the slit being sufficiently narrow, but the lines will still have finite widths. For spectra of atoms or molecules in the gas phase, there are three main causes of line broadening:

1. **Pressure broadening**, due to collisions between the atoms or molecules. This broadening can be minimized by minimizing the pressure.
2. **Doppler broadening**, due to the atoms or molecules travelling in different directions relative to the detector. Removal of Doppler broadening can be achieved, but with some difficulty.
3. **Natural line broadening** is what remains of the line width when the other two sources of broadening have been removed. This broadening is inherent in the spectrum of the atom or molecule, and cannot be removed.

Collisions involve the transfer of energy between the colliding particles, thereby changing, slightly, the energy (or wavelength or wavenumber) at which the transition occurs.

The fact that the whistle of a train increases in frequency as it travels towards an observer, and decreases as it moves away, is an example of the **Doppler effect**. Similarly, an atom or molecule travelling towards a detector absorbs (or emits) radiation at a higher frequency than one travelling away from the detector. The true frequency is either when it has zero velocity, or when it is travelling in a direction perpendicular to the direction of observation.

## 8.4 Spectroscopy in Various Phases

For the most detailed spectroscopic investigation of the structure of atoms and molecules, the **gas phase** is by far the most informative. This is because each atom or molecule, under ideal conditions, behaves independently of its nearest neighbours. These ideal conditions include a gas pressure which is sufficiently low for no collisions to occur during the time taken for a transition between energy levels to take place.

In general, though, much of molecular spectroscopy, particularly for analytical purposes, is carried out in the **liquid phase**. The sample may be the pure liquid, but is often a dilute solution in a solvent which is transparent in the region of the spectrum being investigated.

Spectroscopy of molecules is also carried out in the **solid phase**. This may be a crystalline or amorphous pure solid, but most studies are carried out with the sample embedded in a solid, spectroscopically transparent matrix. The technique is **matrix isolation spectroscopy**. This is usually carried out at very low (liquid nitrogen or liquid helium) temperature, one effect of which is to sharpen the spectrum. The sample is in very low concentration in order to minimize interactions between nearest neighbour sample molecules, thereby imitating a low-pressure gas. Interactions with the matrix are unavoidable and may complicate the spectra; in particular, there are frequency shifts compared to the gas phase spectrum.

An important advantage of matrix isolation spectroscopy is that a short-lived molecule, such as a radical, may be deposited in, or created in, a matrix. At very low temperatures, and under the conditions of isolation in the matrix, the short-lived species remains sufficiently stable for spectra to be obtained.

## Summary of Key Points

**1.** *Main features of the experimental method for obtaining an absorption spectrum*
The principal features are the source of radiation, the absorption cell, a dispersing element (except for microwave spectroscopy) and the detector. Some details are given of those used in microwave, far-, mid- and near-infrared, visible and near- and far-ultraviolet spectroscopy.

**2.** *Absorbance and the Beer–Lambert law*
Definition of absorbance and its relation to the length of the absorption cell and the concentration of the sample.

**3.** *Raman spectroscopy*
The nature of Raman scattering and the method of observation. Use of lasers as ideal sources of radiation.

**4.** *Widths of spectral lines*
Effect of the width of the entrance slit of a spectrometer. Effects of pressure and the Doppler effect on the natural line width in a gas phase sample.

**5.** *Spectra in different phases*
Gas phase spectra are ideal for maximum information on atomic and molecular structure. Liquid phase, especially solution, spectra in the mid- and near-infrared, visible and near-ultraviolet regions are commonly used for analytical purposes. Spectra in the solid phase often employ matrix isolation techniques.

## Problems

**8.1.** In the electronic absorption spectrum, a molecule has a molar absorption coefficient of 10 500 $mol^{-1}$ $dm^3$ $cm^{-1}$ at the wavelength of observation. The sample is in a cell of length 103 mm, and 15.4% of the radiation is transmitted. What is the concentration of the molecule?

**8.2.** Compare the Rayleigh scattering intensity at 11 900 $cm^{-1}$, in the red region, with that at 25 900 $cm^{-1}$, in the blue region of the spectrum.

# 9
# Rotational Spectroscopy

## Aims

The overall aim of this chapter is to understand molecular rotational spectra resulting from transitions between the rotational energy levels discussed in Chapter 7. By the end of this chapter you should be able to:

- Understand the rotational transitions (selection rules, dipole moment and polarizability requirements) that are allowed in microwave, millimetre wave, far-infrared and Raman spectra
- Understand the Boltzmann distribution law relating to rotational and vibrational level populations
- Understand the effects of nuclear spin on rotational level populations
- Appreciate why even spherical top molecules may show a rotational spectrum
- Understand how the bond length may be obtained from $B_0$, the $B$-value in the zero-point level, in a diatomic molecule, and the use of isotopic substitution for obtaining the structure of a polyatomic molecule

## 9.1 Introduction

In Chapter 7, rotational energy levels (term values) for diatomic molecules and linear, symmetric top, spherical top and asymmetric top polyatomic molecules were discussed. In general, the rigid rotor approximation was assumed, but the effects of centrifugal distortion were introduced for diatomic and linear polyatomic molecules.

In the present chapter we shall discuss the transitions between these levels which are observed in rotational spectroscopy. Of prime importance are the **selection rules**: these determine which transitions are **allowed** and which are **forbidden**. In addition, we shall consider the **intensity** of these transitions. The most important contribution to the relative intensity of a transition is from the relative **population** of the initial state. In highly symmetrical molecules, such as homonuclear diatomics, nuclei with non-zero **nuclear spin** also have an important effect on transition intensities.

## 9.2 Rotational Spectroscopy of Diatomic and Linear Polyatomic Molecules

### 9.2.1 Microwave, Millimetre Wave and Far-infrared Spectroscopy

The rotational term values for a diatomic or linear polyatomic molecule were given in equations (7.5) and (7.6) and the energy levels illustrated in Figure 7.2. A similar set of energy levels is shown in Figure 9.1.

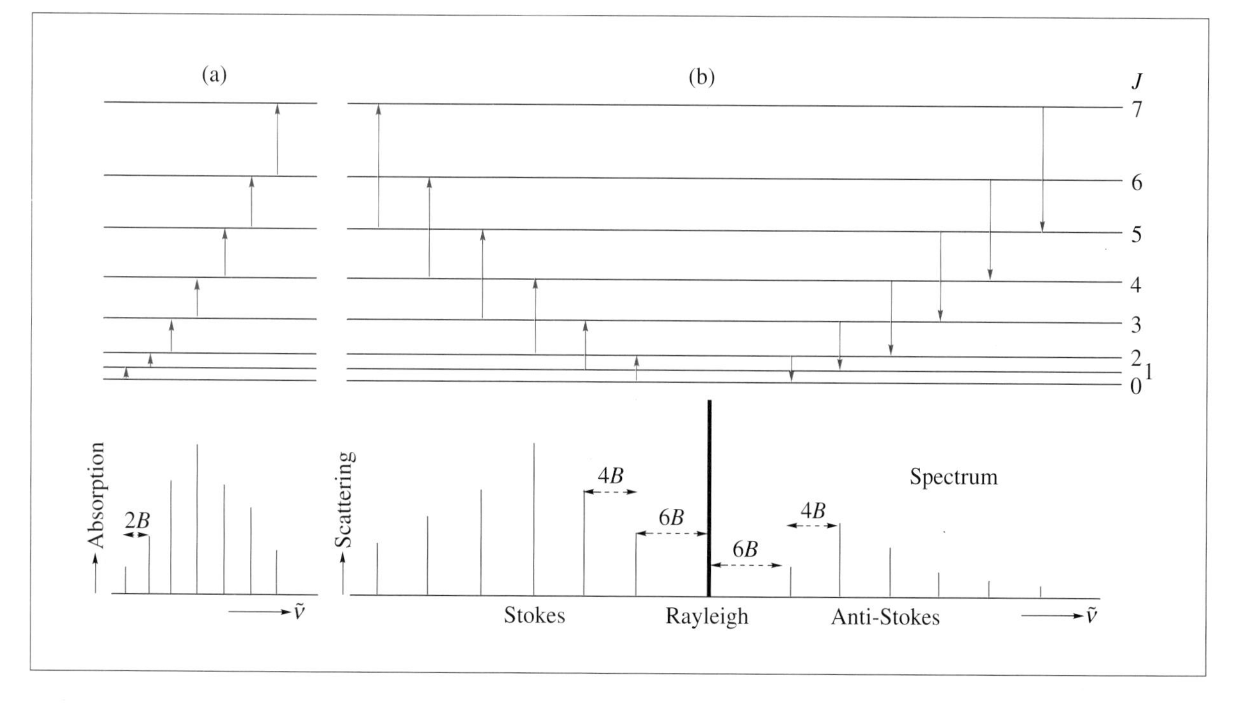


**Figure 9.1** Rotational energy levels of a diatomic or linear polyatomic molecule showing allowed transitions in (a) microwave, millimetre wave or far-infrared and (b) Raman spectroscopy

For rotational transitions between these energy levels that occur in the microwave, millimetre wave or far-infrared region, the selection rule is:

$$\Delta J = J' - J'' = \pm 1 \tag{9.1}$$

The selection rule in equation (9.1) is derived, using quantum mechanics, from the rotational wave functions for the two combining rotational states. In this treatment there is no distinction between upper and lower states, but since $\Delta J$ is defined as $J'–J''$, $\Delta J = -1$ has no meaning. It is often thought, erroneously, that $\Delta J = -1$, or $\Delta N = -1$, where $N$ is any quantum number, refers to an emission process. However, equation (9.1) makes it clear that $\Delta J = +1$ applies to both absorption and emission.

For a diatomic molecule, the **dipole moment**, $p$, is given by $p = Qr$, where $Q$ is the (positive or negative) electrical charge on each of the atoms and $r$ is the bond length. The dipole moment is a measure of the asymmetry of the charge distribution on the two atoms.

where $J'$ and $J''$ are the rotational quantum numbers for the upper and lower state, respectively, of the transition. Such transitions are shown, in absorption, in Figure 9.1(a). The conventional method for labelling a transition is $J'–J''$, so that the transitions in Figure 9.1(a) are 1–0 to 7–6.

In addition to this selection rule, there is a further important condition for there to be a rotational absorption (or emission) spectrum: this is that the molecule must have a non-zero **dipole moment**, $p$. For example, $F_2$, for which $p = 0$, has no such spectrum, while HF, for which $p \neq 0$, does have a rotational spectrum.

Treating the molecule as a rigid rotor and neglecting any effects of centrifugal distortion (see Section 7.2.2), the rotational term values $F(J)$, from equation (7.5) or (7.6), are $BJ(J + 1)$. Applying the selection rule of equation (9.1) gives:

$$F(J + 1) - F(J) = B(J + 1)(J + 2) - BJ(J + 1) = 2BJ + 2B \quad (9.2)$$

where $J$ is understood to be the lower state quantum number, $J''$. The rotational constant $B$ is given by:

$$B = h/8\pi^2 I \quad \text{or} \quad h/8\pi^2 cI \quad (9.3)$$

with units of frequency or wavenumber, respectively; $I$ is the moment of inertia given by equation (7.4).

The result in equation (9.2) shows that the transition with $J = 0$ occurs at a frequency (or wavenumber) of $2B$ and that subsequent transitions occur at intervals of $2B$. These transitions are shown in the energy level diagram in Figure 9.1(a), and the resulting spectrum is shown below the diagram. Whether the transitions are in the far-infrared, millimetre wave or microwave region of the spectrum depends on the value of $B$. Because of the inverse relationship between the moment of inertia and the $B$-value, a heavy molecule, having a large moment of inertia, has a small $B$-value. Consequently, the rotational spectrum appears in the microwave and/or millimetre wave region. For a light molecule, having a large $B$-value, the transitions are more likely to fall in the far-infrared region. Much greater accuracy of measurement of transition frequencies can be achieved in the microwave and millimetre wave regions, where the experimental resolution (see Section 1.2) obtainable is very much higher than in the far infrared.

The most important aspect of the intensity of a rotational transition, or *any* type of spectroscopic transition, is whether it is sufficiently high for the transition to be observed. In any absorption spectrum the most important contribution to the intensity comes from the relative population of the initial state of the transition, *i.e.* what proportion of the molecules are in that state.

**Worked Problem 9.1**

**Q** Treating each molecule as a rigid rotor, calculate the frequency of the 1–0, 3–2 and 5–4 rotational transitions of $^{12}C^{16}O$ ($B$ = 57.64 GHz), $^{1}H^{35}Cl$ ($B$ = 313.0 GHz) and $^{27}Al^{19}F$ ( $B$ = 16.56 GHz) and indicate the region of the spectrum in which they appear.

**A** Using equation (9.2) to calculate the frequencies of the transitions gives:

| *Molecule* | $\nu$(1–0)/GHz | $\nu$(3–2)/GHz | $\nu$(5–4)/GHz |
|---|---|---|---|
| $^{12}C^{16}O$ | 115.3 (mmw)[a] | 345.8 (mmw) | 576.4 (mmw) |
| $^{1}H^{35}Cl$ | 626.0 (fir)[a] | 1878 (fir) | 3130 (fir) |
| $^{27}Al^{19}F$ | 33.12 (mw)[a] | 99.36 (mw/mmw) | 165.6 (mmw) |

[a] mw = microwave, mmw = millimetre wave, fir = far infrared

Although Figure 2.1 indicates that the upper frequency limit of the millimetre wave region is about 600 GHz, sources have been developed which extend into the region of 1 THz ($10^3$ GHz).

Energy level populations are governed, principally, by the **Boltzmann distribution law** which states that the population, $N_n$, of an upper state $n$ is related to the population, $N_m$, of a lower state $m$ by:

$$N_n/N_m = (g_n/g_m)\exp(-\Delta E/kT) \quad (9.4)$$

where $g_m$ and $g_n$ are the degeneracies of the lower and upper states, respectively, $\Delta E$ is the energy separation of the two states, $k$ is the Boltzmann constant and $T$ is the temperature of the sample.

Remember that **degenerate states** are those which have the same energy but different wave functions.

In the absence of an external electric or magnetic field, rotational energy levels of a diatomic molecule are $(2J + 1)$-fold degenerate. A magnetic field, for example, splits each $J$-level into $2J + 1$ components, identified by the **magnetic quantum number**, $M_J$, where:

$$M_J = J, J - 1, \ldots -J \quad (9.5)$$

If we take the lower state, $m$, to be that with $J = 0$, equation (9.4) becomes:

$$N_J/N_0 = (2J + 1)\exp[-hBJ(J + 1)/kT] \quad (9.6)$$

where the rotational term values, in units of frequency, have been converted to energy by multiplying by the Planck constant, $h$. There are four particularly important features of this equation:

1. As $J$ increases, the exponential factor decreases and, because of the exponential relationship, this decrease is rapid.
2. As we go to heavier molecules, $B$ decreases and the decline of the exponential factor becomes less pronounced: the population distribution moves to higher $J$-values.
3. As the temperature increases, the population distribution shifts towards levels with higher $J$-values.
4. Whereas the effect of the negative exponential factor is to impose a smooth, and relatively rapid, decline in population with increasing $J$, the effect of the $(2J + 1)$ degeneracy factor is to tend to reverse this decline owing to the imposition of an increase which is linear. The net effect of the competing negative exponential and linear terms is that the population increases initially as $J$ increases, but eventually the more effective exponential factor wins and the population declines to zero. Since relative populations have the greatest effect on transition intensities, rotational spectra show an intensity distribution which rises to a maximum as $J$ increases, then falls smoothly to zero. This effect is shown in the rotational spectrum in Figure 9.1(a).

### Worked Problem 9.2

**Q** The approximate $B$-values for $^{1}H^{79}Br$ and $^{1}H^{127}I$ are 250 and 193 GHz, respectively. Neglecting centrifugal distortion, calculate the population, relative to that of the $J = 0$ level, of the $J = 10$ level of each molecule at a temperature of (a) 0 °C (273 K) and (b) 100 °C.

**A** (a) For HBr, $J = 10$ and $T = 273$ K:
$hBJ(J + 1)/kT = 6.626 \times 10^{-34}$ J s $\times\ 250 \times 10^{9}$ s$^{-1} \times 10 \times 11/1.381 \times 10^{-23}$ J K$^{-1} \times 273$ K $= 4.83$
$\therefore$ from equation (9.6):
$N_{10}/N_0 = 21 \times \exp(-4.83) = 0.168$
Similarly, for HI, $J = 10$ and $T = 273$ K:
$N_{10}/N_0 = 0.504$
(b) For HBr, $J = 10$ and $T = 373$ K:
$N_{10}/N_0 = 0.609$
For HI, $J = 10$ and $T = 373$ K:
$N_{10}/N_0 = 1.37$

These results show that, for HBr and HI at 273 K, and for HBr at 373 K, $J$ = 10 is sufficiently large for the level to be less highly populated than $J$ = 0; the level has higher energy than that showing maximum population. For HI at 373 K, the population of the $J$ = 10 level is a little larger than that of $J$ = 0. This latter result alone does not tell us whether $J$ = 10 is lower or higher than the level of maximum population, but the result at 273 K indicates that it must be higher.

The general appearance of rotational spectra of linear polyatomic molecules is similar to that of diatomics, showing a series of lines separated by $2B$, in the rigid rotor approximation, and with intensities rising from the $J$ = 1–0 transition to a maximum then falling smoothly to zero. The $B$-value of a linear polyatomic molecule may be very small as a consequence of the length of the chain of atoms and, possibly, the inclusion of heavy atoms. One such example is H–C≡C–C≡C–C≡C–C≡N (hepta-2,4,6-triynenitrile, cyanohexatriyne), for which part of the microwave spectrum is shown in Figure 9.2.

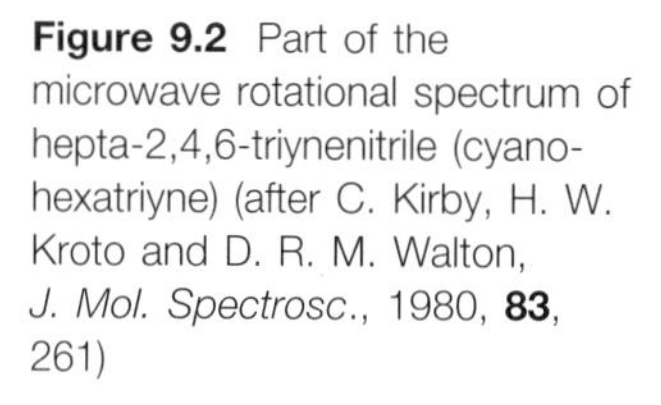
**Figure 9.2** Part of the microwave rotational spectrum of hepta-2,4,6-triynenitrile (cyanohexatriyne) (after C. Kirby, H. W. Kroto and D. R. M. Walton, *J. Mol. Spectrosc.*, 1980, **83**, 261)

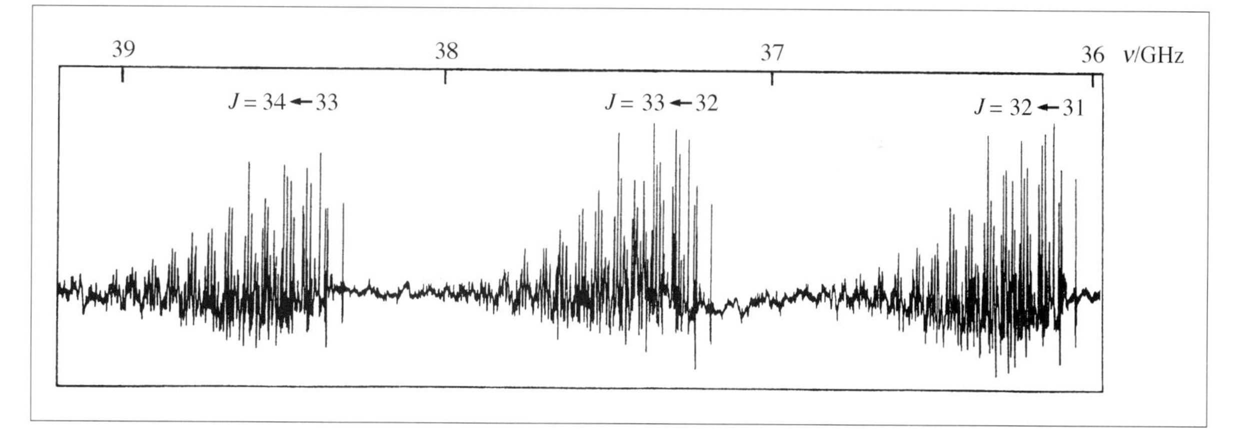


For this long and heavy molecule, $B$ = 564.000 74 MHz, and the rotational spectrum lies in the microwave region. The figure shows that the intensity of the transitions falls with increasing $J$; the intensity maximum occurs at lower $J$. It is clear from the spectrum that there is not just a single line associated with each rotational transition but a complex group of lines. This is characteristic of a polyatomic linear molecule; the reason for this will be discussed in Section 9.3.

The observation of this molecule in the laboratory preceded its detection in the interstellar medium (the region between stars). The latter observation, also from its microwave rotational spectrum, was made possible only by the accurate terrestrial measurement of transition

### 9.2.2 Raman Spectroscopy

In Section 8.2 we have seen that Raman spectroscopy involves a scattering process. Rayleigh scattering is of unchanged wavelength, and will not concern us further. Raman scattering is of increased or decreased wavelength, and is known as Stokes or anti-Stokes Raman scattering, respectively. Raman scattering by molecules may involve transitions between rotational, vibrational or, very rarely, electronic states. Here we consider transitions between rotational states.

For Raman scattered radiation involving rotational transitions, the selection rule is:

$$\Delta J = 0, \pm 2 \tag{9.7}$$

Figure 9.1(b) shows the overall results of transitions with $\Delta J = \pm 2$; those with $\Delta J = 0$ are coincident with the Rayleigh scattering, and are of no further consequence.

In order for scattering to occur, the incident radiation must not be absorbed by the sample molecules. So what does the radiation do to the sample? It **polarizes** it, which means that, to some extent, it tends to separate positive and negative charges, the nuclei and the electrons. The **polarizability** of the molecule is a measure of the degree to which it can be polarized. For a molecule to have a rotational Raman spectrum it is essential that the polarizability is **anisotropic**. All molecules except spherical tops, such as methane, have anisotropic polarizability and show a rotational Raman spectrum. This is an important difference from the more stringent requirement that a molecule must have a dipole moment in order to show a microwave, millimetre wave or far-infrared rotational spectrum. For homonuclear diatomic molecules, such as $N_2$ and $O_2$, and symmetrical linear polyatomic molecules, such as H–C≡C–H and N≡C–C≡N, only Raman spectroscopy can be used for investigation of the rotational spectrum. For molecules having a dipole moment, the advantage of using microwave or millimetre wave spectroscopy is that they are higher resolution techniques, leading to the determination of parameters such as $B$-values with very much higher accuracy.

An **isotropic property** is one which is the same in all directions, such as the electrical conductivity of a liquid, whereas an anisotropic property is one which is different in different directions. An example of an anisotropic property is the electrical conductivity of a crystalline material.

Consider the two $J = 2$–$0$ transitions shown in Figure 9.1(b). That with an increased wavelength (decreased wavenumber), compared to that of the monochromatic (laser) radiation used, is the Stokes transition. Initially, the molecule is in the $J = 0$ level. One quantum of the incident radiation can be considered to take the molecule into an excited **virtual state**, $V_0$. Then, obeying the selection rules in equation (9.7), it may revert to the $J = 0$ level, a transition which is coincident with the Rayleigh scattering, or, in a Stokes transition, to the $J = 2$ level. On the other hand, in the anti-Stokes transition, one quantum of the radiation takes the molecule from the $J = 2$ level to a virtual state, $V_2$. The transition back to $J = 2$ is coincident with the Rayleigh scattering while that to $J = 0$ is an anti-Stokes transition.

A **virtual state** is a state in which the molecule is polarized; it is not a real state (an eigenstate). If it were a real state, the radiation would be absorbed rather than scattered.

Since $\Delta J$ is defined as $J'-J''$, the separations of the $\Delta J = 2$, Stokes and anti-Stokes lines from that of the incident radiation are given, in terms of wavenumber, by:

$$|\Delta\tilde{\nu}| = F(J + 2) - F(J) = B(J + 2)(J + 3) - BJ(J + 1) = 4BJ + 6B \quad (9.8)$$

where $|\Delta\tilde{\nu}|$ is the magnitude, irrespective of the sign, of $\Delta\tilde{\nu}$. Therefore the rotational transitions are observed at wavenumbers given by:

$$\tilde{\nu} = \tilde{\nu}_0 \pm (4BJ + 6B) \quad (9.9)$$

where $\tilde{\nu}_0$ is the wavenumber of the incident laser radiation. Figure 9.1(b) shows the resulting pattern of Stokes and anti-Stokes lines in the spectrum. The first member of each set of lines is separated from the Rayleigh scattered radiation, at $\tilde{\nu}_0$, by $6B$ and subsequent spacings are $4B$.

Figure 9.3 shows a rotational Raman spectrum of $N_2$, which was obtained with 632.8 nm wavelength radiation from a helium–neon laser. This spectrum, and the one illustrated in Figure 9.1(b), both show the rise and fall in intensities which are a consequence of the Boltzmann population distribution among rotational levels in equation (9.6). Each anti-Stokes transition is rather less intense than the equivalent Stokes transition because the initial rotational level of the anti-Stokes transition is at a higher energy and therefore less highly populated.

In addition to the effect on the intensities of initial state populations, the rotational Raman spectrum of $^{14}N_2$ in Figure 9.3 also shows an alternation of the intensities with $J$. Careful measurement of intensities shows that the alternation is in the ratio 2:1. This is a consequence of the lower levels of transitions with even values of $J$ having twice the population of those with odd $J$. This effect is due to nuclear spin, which will now be discussed further.

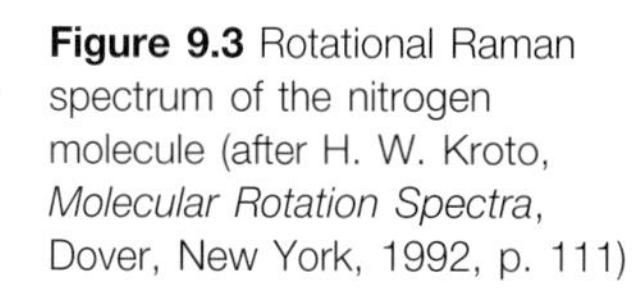
**Figure 9.3** Rotational Raman spectrum of the nitrogen molecule (after H. W. Kroto, *Molecular Rotation Spectra*, Dover, New York, 1992, p. 111)

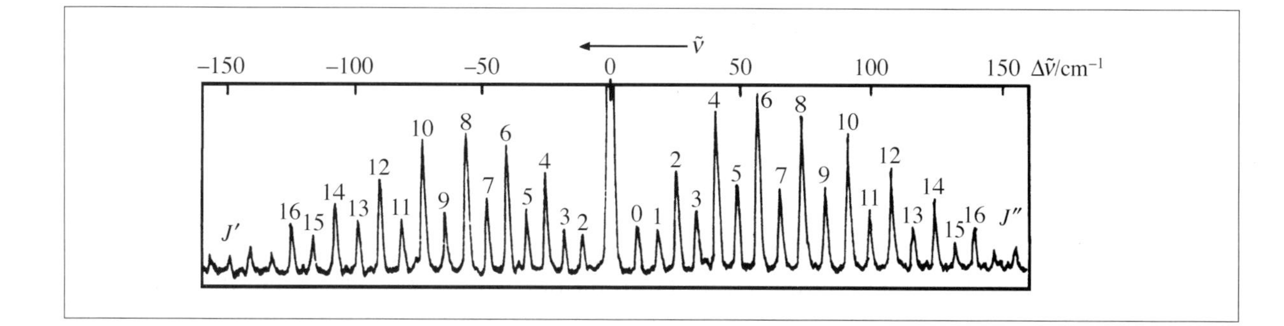

### 9.2.3 Effects of Nuclear Spin

Nuclear spin, and the associated angular momentum, were introduced in Section 3.3 in the context of the hydrogen atom. For example, the nuclear spin quantum number $I = \frac{1}{2}$ for $^1$H and the quantum number $m_I = +\frac{1}{2}$ or $-\frac{1}{2}$ (see equation 3.25).

All nuclei contain protons and neutrons. For each of these, $I = \frac{1}{2}$, and the value of $I$ for the nucleus is the result of the vector sum of all the associated angular momenta. In general, $I$ may be 0, 1, 2, ..., and the nuclei are said to be **Bose particles,** or **bosons,** or it may be $\frac{1}{2}, \frac{3}{2}, \frac{5}{2}$, ..., in which case the nuclei are said to be **Fermi particles** or **fermions.**

If a wave function is symmetric to nuclear exchange, this means that it does not change sign if we interchange the two identical nuclei; if it is antisymmetric, the wave function changes sign.

Consider the two nuclei in $^1H_2$. For each of them, $m_I = \pm\frac{1}{2}$ so that there are four possible combinations: $+\frac{1}{2},+\frac{1}{2}$; $-\frac{1}{2},-\frac{1}{2}$; $+\frac{1}{2},-\frac{1}{2}$; $-\frac{1}{2},+\frac{1}{2}$. Associated with these combinations are four **nuclear spin wave functions,** $\psi_{ns}$, of which *three* are symmetric to nuclear exchange and *one* is antisymmetric. Equation (6.2), giving the breakdown of the total molecular wave function, $\psi_{total}$, into the electronic, vibrational and rotational contributions, did not include the possibility that $I \neq 0$ for one of more of the nuclei. When this is the case, the total wave function becomes:

$$\psi_{total} = \psi_e \psi_v \psi_r \psi_{ns} \tag{9.10}$$

where $\psi_{ns}$ is the wave function due to nuclear spin.

It is only the $\psi_r \psi_{ns}$ part of the wave function which concerns us here. For fermions, such as $^1$H, this part must be antisymmetric to nuclear exchange. For all homonuclear diatomic molecules, $\psi_r$ is symmetric to nuclear exchange for all even $J$ and antisymmetric for all odd $J$. Therefore, in $^1H_2$, the single antisymmetric $\psi_{ns}$ combines only with $\psi_r$ for even $J$, and the three symmetric $\psi_{ns}$ with $\psi_r$ for odd $J$. The result is that there is a nuclear spin **statistical weight** alternation of 1:3 for $J$ even:odd. This alternation is reflected in any rotational spectrum of $^1H_2$ (or $^{19}F_2$, since $I = \frac{1}{2}$ for $^{19}$F) showing an intensity alternation of 1:3 for the lower state $J$-value even:odd, superimposed on the rise and fall in intensity due to the Boltzmann population distribution.

For the $^{14}$N nucleus $I = 1$, and $m_I = 1$, 0 or $-1$. For the two nuclei in $^{14}N_2$ there are nine possible combinations of the two values of $m_I$ and nine associated $\psi_{ns}$ wave functions. Three of these are symmetric and six are antisymmetric to nuclear exchange. Because $^{14}$N is a boson, $\psi_r \psi_{ns}$ must be symmetric and therefore the nuclear spin statistical weights of the rotational levels are 6:3 for $J$ even:odd. The resulting intensity alternation in the rotational Raman spectrum is clear from Figure 9.3. There is an identical nuclear spin statistical weight alternation in the populations of rotational levels of $^2H_2$, since $I = 1$ for $^2$H.

Nuclear spin statistical weights for the $J = 0$–3 levels of some diatomic molecules are summarized in Figure 9.4.

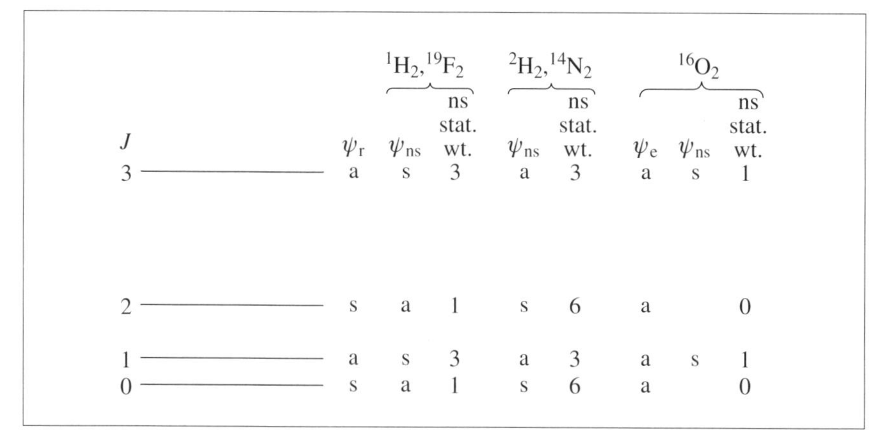


**Figure 9.4** Nuclear spin statistical weights for the $J = 0$ to 3 rotational energy levels of some homonuclear diatomic molecules

## Box 9.1 Nuclear Spin Statistical Weights

For a nucleus with spin quantum number $I$, there are $(2I + 1) \times (I + 1)$ nuclear spin wave functions which are symmetric to nuclear exchange, and $(2I + 1)I$ which are antisymmetric. Therefore, for $^1H_2$ and $^{19}F_2$ ($I = \frac{1}{2}$), $^2H_2$ and $^{14}N_2$ ($I = 1$) and $^{16}O_2$ ($I = 0$) we have:

| | $^1H_2$, $^{19}F_2$ | $^2H_2$, $^{14}N_2$ | $^{16}O_2$ |
|---|---|---|---|
| $(2I + 1)(I + 1)$ symmetric wave functions | 3 | 6 | 1 |
| $(2I + 1)I$ antisymmetric wave functions | 1 | 3 | 0 |

Note that $^{16}O_2$ has no antisymmetric nuclear spin wave functions. There is a further complication in that the electronic ground state of $O_2$ is a triplet state (see Section 5.1.2) in which there are two parallel electron spins. The result of this is that $\psi_e$, the electronic wave function, is antisymmetric to nuclear exchange. (We have been able to ignore $\psi_e$ in the other molecules since they have singlet ground states for which $\psi_e$ is symmetric to nuclear exchange.) The nuclei are bosons, from which it follows that $\psi_e\psi_r\psi_{ns}$ must be symmetric to nuclear exchange. Thus, the rotational levels with even $J$-values have zero statistical weight and therefore zero population: the levels are absent and no transitions involving them as initial levels are observed. Only the levels with odd $J$-values are populated.

### 9.2.4 Spectra of Molecules in Vibrationally Excited States

Vibrational energy levels of diatomic molecules have been discussed in Section 6.2 and illustrated, for an anharmonic oscillator, in Figure 6.3. The relative populations of these vibrational levels are determined by the Boltzmann distribution law of equation (9.4), in the same way as the populations of rotational levels. For the population of a vibrational energy level with quantum number $v$, relative to that of the zero-point level with $v = 0$, equation (9.4) becomes:

$$N_v/N_0 = \exp\{-hc[G(v) - G(0)]/kT\} \tag{9.11}$$

where the vibrational term values $G(v)$ are given by equation (6.12), for a harmonic oscillator, or equation (6.13), for an anharmonic oscillator; the vibration is non-degenerate ($g_0 = g_v = 1$). This relationship shows that the population of, say, the $v = 1$ level is very small for a light molecule, with a large vibration wavenumber $\omega$, compared to a heavy molecule with a relatively small value of $\omega$. It also shows that the population of this level can be increased by increasing the temperature, but, because of the exponential nature of the relationship, it requires a very large increase in temperature to increase the population appreciably.

**Worked Problem 9.3**

**Q** The vibration wavenumbers for $I_2$ and $H_2$ are 215 and 4400 cm$^{-1}$, respectively. Using equation (9.11), calculate the population ratio $N_1/N_0$ for both molecules at temperatures of 20.0 °C and 1000 °C. Note that 0 °C = 273.2 K.

**A** $hc/kT = 6.626 \times 10^{-34}$ J s $\times$ $2.998 \times 10^{10}$ cm s$^{-1}$/$1.381 \times 10^{-23}$ J K$^{-1}$ $\times$ 293.2 K

$= 4.906 \times 10^{-3}$ cm

For $I_2$, $G(1) - G(0) = 215$ cm$^{-1}$

$\therefore$ $hc[G(1) - G(0)]/kT = 1.055$

and $\exp\{-hc[G(1) - G(0)]/kT\} = 0.348$

Similar calculations for $T = 1273.2$ K, and for $H_2$, are summarized below:

| | $N_1/N_0$ at 20.0 °C | $N_1/N_0$ at 1000 °C |
|---|---|---|
| $I_2$ | 0.348 | 0.734 |
| $H_2$ | $4.20 \times 10^{-10}$ | $1.78 \times 10^{-3}$ |

These results illustrate the large differences between the two molecules. There are virtually no $H_2$ molecules in the $v = 1$ level at 20.0 °C, and even at 1000 °C there are very few. For $I_2$, the number of molecules in the $v = 1$ (and even in the $v = 2$) level is very significant.

Figure 6.9 shows that there are rotational energy levels associated with each vibrational level. If the molecule has a sufficiently low vibration wavenumber, $\omega$, any rotational spectrum may show rotational transitions within the $v = 1$, as well as within the $v = 0$, level. These additional rotational transitions obey the same selection rules of equation (9.1) or (9.7) for microwave/infrared or Raman spectra, respectively. However, as will be discussed in more detail in Chapter 10, the rotational constant, $B$, varies with the vibrational energy level in which the rotational energy levels occur. Consequently, the rotational transitions in the $v = 1$ level are slightly displaced from those in the $v = 0$ level, and they are weaker because of the population difference.

A linear polyatomic molecule has $3N - 5$ vibrational modes (see Section 6.3), where $N$ is the number of atoms. Some of these modes may have low vibration wavenumbers and may be appreciably populated, resulting in groups of rotational transitions accompanying those with all $v = 0$. The microwave spectrum of the linear molecule hepta-2,4,6-triynenitrile (cyanohexatriyne) (Figure 9.2) shows such groups of lines. Although the molecule is linear, it is extremely flexible about the various atoms of the chain. The large number of lines in each group is due to the resulting low-lying energy levels involving the various bending motions.

## 9.3 Microwave, Millimetre Wave and Far-infrared Spectroscopy of Non-linear Polyatomic Molecules

### 9.3.1 Symmetric Top Molecules

Symmetric top molecules, both prolate and oblate, were introduced in Section 7.3.2. Typical rotational energy levels are shown in Figure 7.5(a) for a prolate, and in Figure 7.5(b) for an oblate, symmetric top.

For a symmetric top molecule to show a microwave, millimetere wave or far-infared rotational spectrum it must have a non-zero dipole moment. Therefore, the prolate symmetric top fluoromethane (Figure 7.3a) shows a rotational spectrum, while benzene, an oblate symmetric top (Figure 7.3b), does not.

To show a pure rotational Raman spectrum the molecule must have an anisotropic polarizability (see Section 9.2.2). All symmetric top molecules have rotational Raman spectra, but they are seldom encountered and will not be discussed further.

The rotational term values for prolate and oblate symmetric top molecules are given in equations (7.11) and (7.12), respectively. The selection rules are:

$$\Delta J = \pm 1 \quad \text{and} \quad \Delta K = 0 \tag{9.12}$$

and the molecule must have a permanent dipole moment. For both prolate and oblate symmetric tops (equations 7.11 and 7.12):

$$F(J+1,K) - F(J,K) = 2BJ + 2B \tag{9.13}$$

These separations of adjacent transitions are independent of $K$, and are the same as those given in equation (9.2) for a diatomic or linear polyatomic molecule.

Figure 9.5 shows part of the far-infrared rotational spectrum of the prolate symmetric top $C^1H_3{}^2H$ in which the $C$–$^2H$ bond lies along the $a$-axis of inertia. The spectrum shows seven transitions, from $J$ = 6–5 to 12–11, with separations of $2B$.

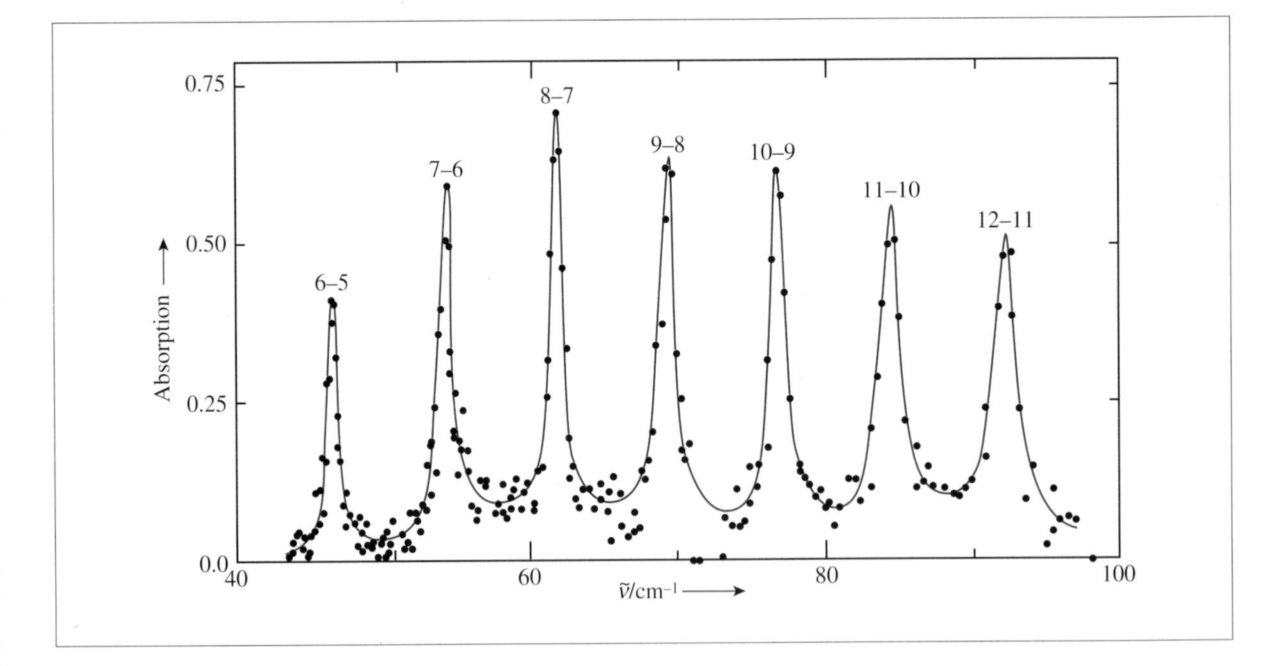


**Figure 9.5** Part of the far-infrared rotational spectrum of $C^1H_3{}^2H$ (monodeuteromethane) (after I. Ozier, W. Ho and G. Birnbaum, *J. Chem. Phys.*, 1969, **51**, 4872)

The most remarkable feature of this spectrum is that it has been observed at all. $C^1H_3{}^2H$ has an extremely small dipole moment of only $5.68 \times 10^{-3}$ D, compared to, say, HF which has a dipole moment of 1.91 D. This results in a very weak absorption spectrum which is difficult to observe.

The resolution in this weak spectrum is rather low so that any fine structure associated with each transition, such as that shown in Figure 9.2, is not observed. There might be fine structure due to rotational transitions in vibrationally excited states, although this molecule does not have vibrations with sufficiently low wavenumbers to be appreciably populated. A more important source of fine structure results from centrifugal distortion, a factor which has been neglected in equations (7.11) and (7.12) and, consequently, in equation (9.13). We shall not consider this effect further except to note that it has the effect of displacing, slightly, transitions with differing values of $K$ and constant $J$.

The **debye** (D) unit is still commonly used as a measure of dipole moment. It is related to the SI unit of coulomb metre by 1 D = 3.335 64 × $10^{-30}$ C m.

### 9.3.2 Spherical Top Molecules

Bearing in mind that a molecule must have a non-zero dipole moment in order to show a microwave or far-infrared rotational spectrum, we would not expect a spherical top such as methane ($CH_4$) or silane ($SiH_4$) to show such a spectrum. However, rotation about one of the C–H or Si–H bonds causes the molecule to distort very slightly, and produces an extremely small dipole moment. For example, the dipole moment produced in $SiH_4$ is only $8.3 \times 10^{-6}$ D; the resulting rotational spectrum is weak but the far-infrared spectrum has been observed.

Only spherical top molecules have isotropic polarizability (polarizability which has the same magnitude in all directions). Therefore, they do not show a rotational Raman spectrum.

### 9.3.3 Asymmetric Top Molecules

The majority of molecules are asymmetric tops, all three principal moments of inertia being unequal (see Section 7.3.4). Microwave and millimetre wave spectra of such molecules are therefore of great importance. However, the complexity of the rotational energy levels and transitions takes the subject beyond the scope of this book.

## 9.4 Molecular Structure Determination

The accurate determination of molecular structure, bond lengths and bond angles is one of the principal end-products of high-resolution (gas phase) spectroscopy. In a diatomic molecule, this involves using a precise value of the rotational constant $B$ to determine the moment of inertia $I$ from the relationships in equation (9.3). Then the bond length $r$ follows from equation (7.2) using the reduced mass $\mu$ from equation (7.3).

However, we have seen that the value of $B$ varies with the vibrational quantum number $v$. Unless we are dealing with rotational spectra in a vibrationally excited state, the value determined will be $B_0$, the value of $B$ in the zero-point vibrational state, and hence the bond length $r_0$. The potential curve shown in Figure 6.3 indicates that what we would really like to determine is $B_e$, and hence $r_e$, corresponding to the bottom of the curve where no vibration is taking place. We shall return to this problem in Chapter 10.

For polyatomic molecules the determination of bond lengths and angles is more complex. For example, as equation (9.13) indicates, the rotational spectrum of a symmetric top molecule produces the value of $B_0$ only, and not $A_0$ or $C_0$. Isotopic substitution, and the assumption that the structure does not change with such substitution, is necessary to obtain more than one geometrical parameter.

For a planar molecule, the sum of the two in-plane moments of inertia is equal to the out-of-plane moment of inertia, *i.e.* $I_a + I_b = I_c$. Consequently, only two moments of inertia, and therefore two rotational constants, are independent. For a non-planar molecule, all three moments inertia, and therefore all three rotational constants, are independent.

Rotational spectra of asymmetric top molecules are more productive. Analysis of the spectrum of a planar molecule results in two rotational constants which can give two geometrical parameters; the spectrum of a non-planar molecule gives three rotational constants.

## Summary of Key Points

**1.** *Rotational selection rules applied to microwave and far-infrared rotational spectra*
For diatomic and linear polyatomic molecules the selection rule is $\Delta J = \pm 1$, and the molecule must have a non-zero dipole moment. For a symmetric top the selection rule $\Delta K = 0$ also applies, and a non-zero dipole moment is essential.

**2.** *Rotational selection rules applied to rotational Raman spectra*
It is only for diatomic, and small linear polyatomic, molecules that rotational Raman spectra are important. The selection rule is $\Delta J = 0, \pm 2$, and the molecule must have an anisotropic polarizability; this includes homonuclear diatomic molecules.

**3.** *Boltzmann distribution law applied to level populations*
Variation of rotational level populations causes the rise and fall of rotational transition intensities as $J$ increases. Application of the law to vibrational level populations can result in appreciable population of levels with $v = 1$, or even higher. Rotational transitions may be observed within these higher vibrational levels.

**4.** *Effects of nuclear spin*
In homonuclear diatomic and symmetrical linear polyatomic molecules, the presence of pairs of symmetrically identical nuclei gives rise to an alternation, with $J$ even or odd, of rotational level populations. Consequently, there is an alternation of intensities in the rotational spectrum.

5. *Determination of molecular structure*
For a diatomic molecule, determination of the rotational constant $B_0$, obtained from transitions within the zero-point vibrational level, gives the bond length $r_0$. In a polyatomic molecule, isotopic substitution may be used to determine more than one geometrical parameter.

## Problems

**9.1.** Make measurements on the rotational Raman spectrum in Figure 9.3 to obtain the best estimate of the *B*-value for $^{14}N_2$.

**9.2.** Given that $m(^{14}N) = 14.00$ u and 1 u $= 1.661 \times 10^{-27}$ kg, determine the bond length of $^{14}N_2$ from the *B*-value obtained in Problem 9.1.

**9.3.** From Figure 9.5, estimate the value of $B$ for $C^1H_3{}^2H$.

# 10 Vibrational Spectroscopy

**Aims**

Vibrational energy levels in diatomic and polyatomic molecules were discussed in Chapter 6. By the end of the present chapter you should be able to:

- Appreciate the fact that there are rotational energy levels associated with all vibrational levels
- Understand the vibrational, and rovibrational, selection rules which apply to infrared and Raman spectra of diatomic and polyatomic molecules
- Use these spectra to obtain rotational constants and structural information
- Understand the mutual exclusion rule relating to vibrations of molecules with a centre of symmetry
- Understand the use of rotational band contours to aid the assignment of infrared spectra of asymmetric top molecules
- Appreciate the use of group vibrations in infrared and Raman spectroscopy when used as analytical tools

## 10.1 Introduction

Vibrational motion in diatomic and polyatomic molecules has been discussed in Chapter 6. The only vibration of a diatomic molecule can be treated in the harmonic oscillator approximation or, more accurately, as that of an anharmonic oscillator. Linear and non-linear polyatomic molecules, with $N$ atoms, have $3N - 5$ and $3N - 6$ vibrations, respectively (see Section 6.3). To a fairly good approximation, each vibration can be treated independently, and that is what we shall do here.

Vibrational spectroscopy, whether infrared or Raman, can be divided into two general types. When vibrational spectra of diatomic and small polyatomic molecules are obtained at high resolution in the gas phase, rotational fine structure may be resolved and interpreted to give important structural information, namely bond lengths and bond angles. On the other hand, it is much more difficult to resolve the rotational fine structure for large molecules, and to obtain detailed structural information. In addition, it may be impracticable to get the molecule into the gas phase with sufficiently high vapour pressure for a spectrum to be obtained. For these molecules, vibrational spectroscopy is commonly used as an analytical tool. For example, it may be used to identify a molecule, or to confirm that a particular group has been substituted as a result of a chemical reaction. Then, it is much more convenient for the sample to be in the form of a pure liquid, a solution or, for an infrared spectrum, a compressed powdered solid in a spectroscopically transparent material.

## 10.2 Infrared Spectra of Diatomic Molecules

The potential energy curve, and the associated vibrational levels, for a diatomic molecule are illustrated in Figure 6.3. Treating the molecule in the harmonic oscillator approximation (see Section 6.2.1) gives the vibrational selection rule:

$$\Delta v = \pm 1 \tag{10.1}$$

In an absorption spectrum, which is the only type that we shall consider here, the most prominent absorption will be from $v = 0$ to $v = 1$. The reason for this is that the population of the $v = 0$ level is higher than that of any other level (see the application of the Boltzmann distribution law to vibrational energy levels in equation 9.11).

When the molecule is treated more accurately as an anharmonic oscillator (see Section 6.2.2), the selection rule relaxes to allow, in principle, all vibrational transitions:

$$\Delta v = \pm 1, \pm 2, \pm 3, \ldots \tag{10.2}$$

In practice, though, the harmonic oscillator approximation is sufficiently good that transitions with $\Delta v = \pm 2, \pm 3, \ldots$, the so-called vibrational **overtones**, are usually very weak.

The potential energy curves shown in Figure 6.9 indicate that there are rotational energy levels associated with all vibrational levels. Figure 10.1 shows some of the rotational levels within the $v = 0$ and 1 levels, and the **rovibrational transitions** which are allowed between them. The

The general requirement for a molecule to have an infrared vibrational spectrum is that there must be a change of dipole moment during the vibrational motion. For a diatomic molecule, this means that it must have a permanent dipole moment and therefore must be heteronuclear; for example, $N_2$ has no dipole moment at any stage during vibration.

vibrational selection rule allows the $v$ = 1–0 transition *provided the molecule has a non-zero dipole moment.*

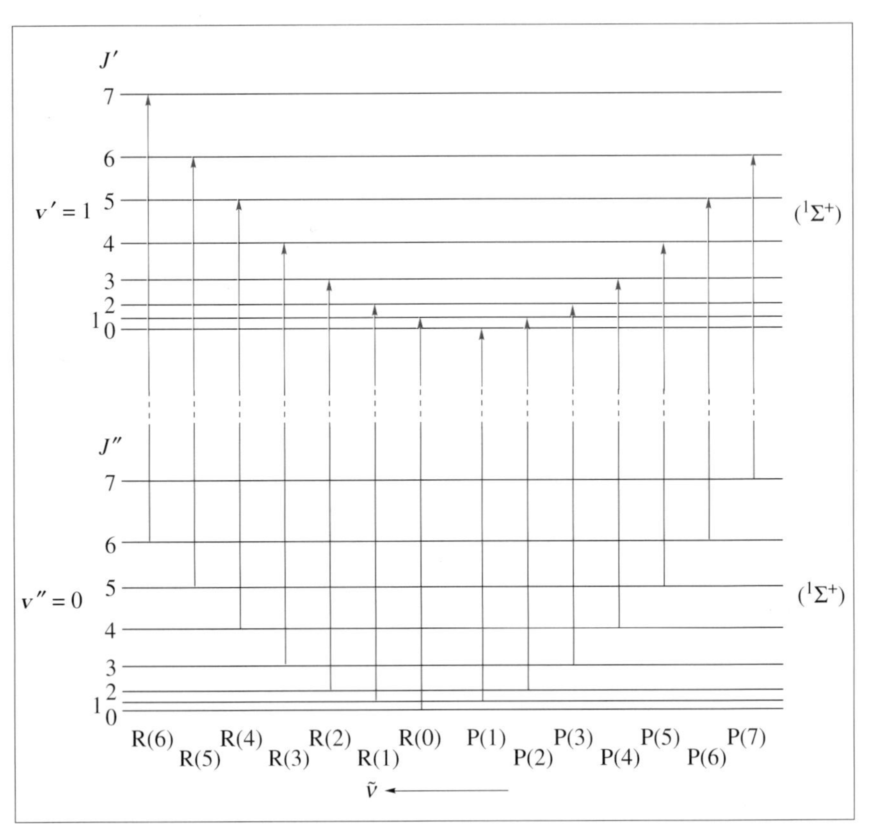


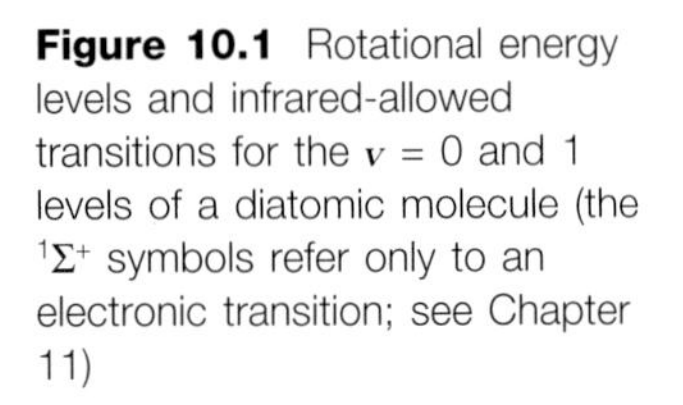

**Figure 10.1** Rotational energy levels and infrared-allowed transitions for the $v$ = 0 and 1 levels of a diatomic molecule (the ${}^1\Sigma^+$ symbols refer only to an electronic transition; see Chapter 11)

The rovibrational transitions in the infrared spectrum obey the selection rule:

$$\Delta J = \pm 1 \tag{10.3}$$

Branches with $\Delta J$ = ... +2, +1, 0, −1, −2 ... are labelled ... S, R, Q, P, O ..., respectively.

The transitions with $\Delta J$ = +1 and −1 constitute the **R and P branch**, respectively, and are labelled R($J$) and P($J$), where $J$ is understood to be the lower state quantum number $J''$.

It was mentioned in Section 9.2.4 that the rotational constant, $B$, varies with the vibrational quantum number, $v$. If at this stage we neglect this small variation, and also centrifugal distortion, the wavenumbers, $\tilde{\nu}[\mathrm{R}(J)]$, of the R-branch transitions are given by:

$$\tilde{\nu}[\mathrm{R}(J)] = \omega_0 + BJ(J+1)(J+2) - BJ(J+1) = \omega_0 + 2BJ + 2B \tag{10.4}$$

where $\omega_0$ is the wavenumber separation of the $J$ = 0 levels and $B$ is the rotational constant, assumed to be the same for both vibrational levels.

Similarly, the wavenumbers, $\tilde{\nu}[\mathrm{P}(J)]$, for the P-branch transitions are given by:

$$\tilde{\nu}[\mathrm{P}(J)] = \omega_0 + B(J-1)J - BJ(J+1) = \omega_0 - 2BJ \qquad (10.5)$$

Note that $\omega_0$, the fundamental vibration wavenumber, cannot be determined directly since all $\Delta J = 0$ transitions, which would comprise a Q-branch, are forbidden.

Equations (10.4) and (10.5) show that, within the limitations of the approximations we have used, the separations within both the R and P branches are $2B$, but the first R- and P-branch lines are separated by $2B$ from the position of the forbidden $J = 0$ to $J = 0$ transition: there is a **zero gap**, as it is called, of $4B$ in the band centre, as shown in Figure 10.1.

The rotational fine structure of the overlapping $v = 1$–$0$ bands of $^1H^{35}Cl$ and $^1H^{37}Cl$ are shown in Figure 10.2. As shown by equations (10.4) and (10.5), this structure is approximately symmetrical about the band centre. The rise and fall of the intensity along both the R- and P-branches reflects the Boltzmann populations of the rotational levels for the $v = 0$ level, as discussed in Section 9.2.1. The relative intensity of the two overlapping bands reflects the natural abundance ratio of 3:1 of the $^{35}Cl$ and $^{37}Cl$ isotopes. The $^1H^{37}Cl$ band is slightly displaced to low wavenumber owing to the larger reduced mass (see equation 6.9).

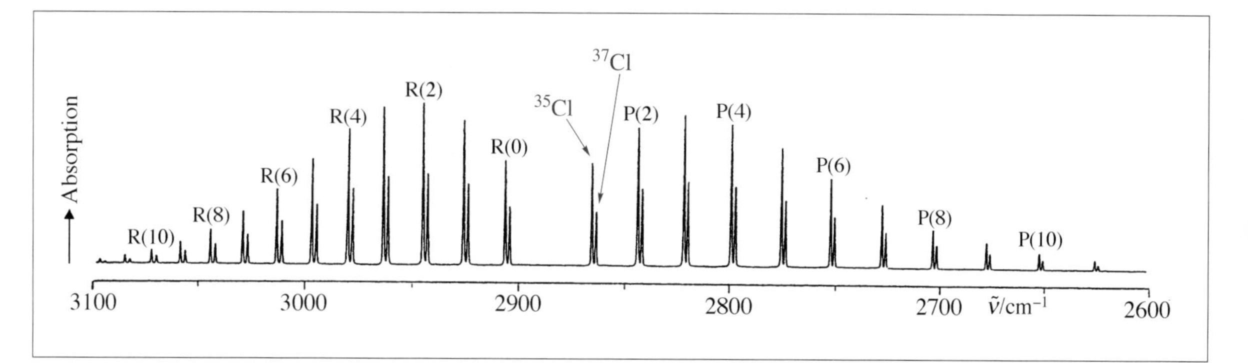


**Figure 10.2** Rotational fine structure of the $v = 1$–$0$ infrared spectrum of HCl

## Worked Problem 10.1

**Q** Use the separation of the R(2) and P(3) lines of $^1H^{35}Cl$ in Figure 10.2 to obtain an estimate of the bond length, given that $m(^1\mathrm{H}) = 1.01$ u and $m(^{35}\mathrm{Cl}) = 34.97$ u. Why would it not be beneficial to use the separation of lines with larger $J$-values?

**A** Convert the distance between the R(2) and P(3) lines to cm$^{-1}$, using the scale on Figure 10.2. Equations (10.4) and (10.5) show that the distance is $12B$.

$\therefore\ 12B = 123\ \text{cm}^{-1}$

$\therefore\ B = 10.25\ \text{cm}^{-1}$ (retaining four figures)

However, $B = h/8\pi^2 c\mu r^2$, from equations (7.2) and (9.3).

The reduced mass $\mu = [1.01 \times 34.97/(1.01 + 34.97)] \times 1.661 \times 10^{-27}$ kg
$= 1.631 \times 10^{-27}$ kg

$\therefore\ r^2 = h/8\pi^2 c\mu B$
$= 6.626 \times 10^{-34}\ \text{J s}/8\pi^2 \times 2.998 \times 10^{10}\ \text{cm s}^{-1} \times 1.631 \times 10^{-27}\ \text{kg} \times 10.25\ \text{cm}^{-1}$
$= 1.674 \times 10^{-20}\ \text{m}^2$

$\therefore\ r = 1.29 \times 10^{-10}\ \text{m} = 1.29$ Å

Only three-figure accuracy is justified in the answer. This value compares well with the accepted value of 1.27 Å. It is only approximate because $B$ has been assumed to be the same in both vibrational states and centrifugal distortion has been neglected. It would not be beneficial to use lines with higher $J$-values because the validity of these approximations decreases with increasing $J$.

Close inspection of the spectrum in Figure 10.2 shows that the separations between adjacent R- and P-branch lines are not constant: the R-branch lines converge, and the P-branch lines diverge, with increasing $J$. Although centrifugal distortion does cause some irregularity in the band structure, the fact that $B_0$ and $B_1$, the values for the $v = 0$ and 1 levels, are not equal has a much larger effect. The rotational constant, $B_v$, varies with $v$ according to:

$$B_v = B_e - \alpha(v + \tfrac{1}{2}) \qquad (10.6)$$

where $B_e$ is the rotational constant corresponding to the equilibrium internuclear distance at the bottom of the potential energy curve in Figure 6.3. $\alpha$ is the **vibration–rotation interaction constant,** which is positive for all diatomic molecules. To determine $B_e$, therefore, it is necessary to determine $\alpha$. This requires two values of $B_v$, for example $B_0$ and $B_1$.

To determine $B_1$ we choose pairs of transitions which have a common *lower* state, for example R(1) and P(1). Because both transitions originate with $J'' = 1$ their wavenumber separation, $\tilde{\nu}[\text{R}(1)] - \tilde{\nu}[\text{P}(1)]$, must depend only on the value of $B_1$. In general:

$$\tilde{\nu}[\text{R}(J)] - \tilde{\nu}[\text{P}(J)] = \omega_0 + B_1(J+1)(J+2) - B_0 J(J+1) - [\omega_0 + B_1(J-1)J - B_0 J(J+1)] = 4B_1(J + \tfrac{1}{2}) \qquad (10.7)$$

Obtaining as many values as possible of $\tilde{\nu}[R(J)] - \tilde{\nu}[P(J)]$ from the spectrum and plotting these against $J + \frac{1}{2}$ gives a straight line with a slope of $4B_1$. Any curvature of the graph, particularly apparent at high $J$, is due to centrifugal distortion.

Similarly, to determine $B_0$ we can take wavenumber differences of pairs of transitions, $\tilde{\nu}[R(J-1)] - \tilde{\nu}[P(J+1)]$, which have an *upper* state in common; for example, R(2) and P(4) both have $J' = 3$ in common. Such differences must depend only on $B_0$:

For a transition between state A and state B, in this case vibrational states, the method of using the separation of pairs of transitions with state A in common to determine the parameters relating to state B, and *vice versa*, is known as the method of **combination differences**.

$$\tilde{\nu}[R(J-1)] - \tilde{\nu}[P(J+1)] = \omega_0 + B_1 J(J+1) - B_0(J-1)J - [\omega_0 + B_1 J(J+1) - B_0(J+1)(J+2] = 4B_0(J+\tfrac{1}{2}) \quad (10.8)$$

Plotting values of $\tilde{\nu}[R(J-1)] - \tilde{\nu}[P(J+1)]$ against $J + \frac{1}{2}$ gives a straight line (except for slight curvature at high $J$) with a slope of $4B_0$.

With these values of $B_0$ and $B_1$, equation (10.6) can be used to give $\alpha$ and $B_e$. From $B_e$, the equilibrium bond length, $r_e$, can be calculated. As discussed in Section 9.4, this is the most important bond length, not least because, since it corresponds to the bottom of the potential energy curve where no vibration is taking place, it is independent of isotopic substitution. For example, $r_e$ is the same for $^{2}H^{35}Cl$ as for $^{1}H^{35}Cl$.

## 10.3 Raman Spectra of Diatomic Molecules

Raman scattering by molecules is very weak, and scattering involving vibrational transitions is particularly weak. Consequently, vibrational Raman spectroscopy in the gas phase, and with sufficiently high resolution to observe rotational fine structure, is not a commonly used technique. However, it is important for homonuclear diatomic molecules, which have no infrared vibrational spectrum. For a vibration to be allowed in the Raman spectrum, there must be a change of polarizability accompanying the vibration: this is the case for *all* diatomic molecules.

The vibrational selection rule for Raman scattering is the same as for an infrared process (see equations 10.1 and 10.2). The $v = 1$–0 transition is dominant, but this may occur as a Stokes or anti-Stokes process, as shown in Figure 10.3. In the Stokes process, the intense, monochromatic laser radiation takes the molecule from the $v = 0$ state to a virtual state, $V_0$, from which it falls back to the $v = 1$ state. Similarly, in the anti-Stokes process, the virtual state $V_1$ is involved in the overall transfer of the molecule from the $v = 1$ to the $v = 0$ state. The Stokes and anti-Stokes transitions lie to low and high wavenumber, respectively, of the exciting radiation. The intensity of the anti-Stokes, relative to that of the Stokes, transition is very low because of the lower population of the $v = 1$, compared to that of the $v = 0$, state. Consequently, vibrational Raman spectroscopy is concerned, usually, only with the Stokes transitions.

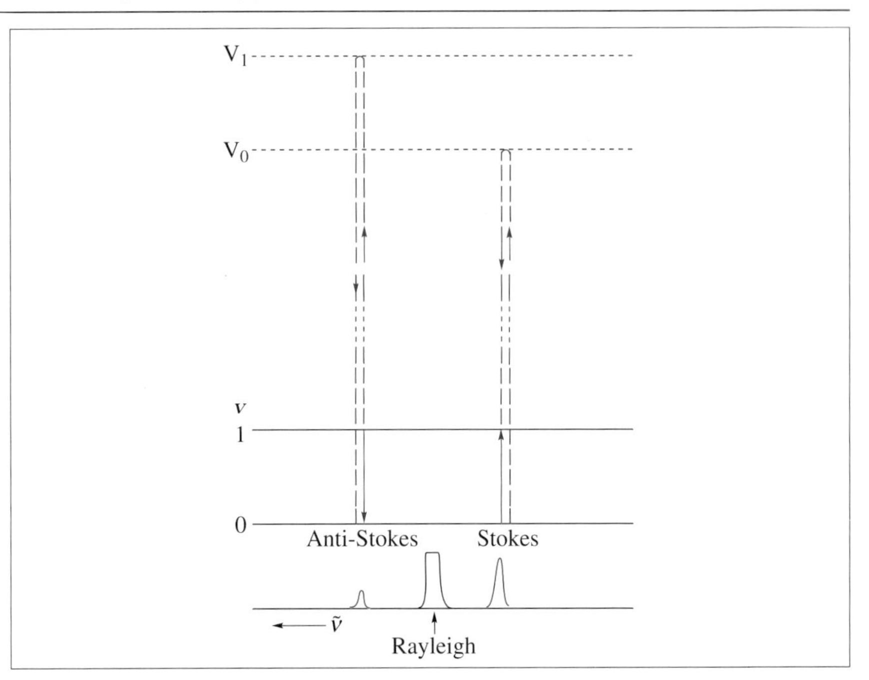


**Figure 10.3** Mechanisms for Stokes and anti-Stokes vibrational transitions in the Raman spectrum of a diatomic molecule

If the resolution is sufficiently high, each vibrational transition shows rotational fine structure. The rotational selection rule is:

$$\Delta J = 0, \pm 2 \tag{10.9}$$

giving rise to a central Q branch ($\Delta J = 0$), and an S ($\Delta J = +2$) and an O ($\Delta J = -2$) branch to high and low wavenumber, respectively. In the approximation that $B_0 = B_1 = B$, all the Q-branch lines are coincident. The first S-branch transition, S(0), is $6B$ to high wavenumber with subsequent S-branch transitions having spacings of $4B$. The first O-branch transition is $6B$ to low wavenumber of the Q-branch, with subsequent spacings of $4B$.

## 10.4 Infrared and Raman Spectra of Linear Polyatomic Molecules

Figure 6.6 shows the normal vibrations of the linear molecule ethyne (acetylene). To see which vibrations are allowed in the infrared spectrum (infrared active) is relatively straightforward. We look for those vibrations during which there is a change of dipole moment. Ethyne has no permanent dipole moment but, when it vibrates in the $\nu_3$ mode, the asymmetric C–H stretching vibration, a dipole moment results from the unequal C–H bonds. Similarly, a dipole moment is produced during the *cis* bending vibration, $\nu_5$. Therefore, both $\nu_3$ and $\nu_5$ are infrared active, as indicated on the left-hand side of the figure.

To determine whether a vibration is Raman active is not so straightforward. Polarizability can be imagined as a three-dimensional ellipsoid centred on the centre of the molecule, as shown in Figure 10.4. A cross section in the plane of the figure is elliptical but, for the polarizability of *any* linear molecule, a cross section in a plane perpendicular to the figure is circular. When the molecule vibrates in the $\nu_1$ mode, the symmetric C–H stretching vibration, we can imagine the polarizability ellipsoid breathing in and out: the polarizability is changing, and the vibration is Raman active. Similarly, $\nu_2$, the C–C stretching vibration, is Raman active. Although this is not so easy to see, $\nu_3$ is not Raman active because the effect on the ellipsoid of the stretching of one C–H bond is cancelled by the contraction of the other one. During the *trans* bending vibration, $\nu_4$, the polarizability ellipsoid is distorted, and the vibration is Raman active. The distortions produced by the motions of the carbon and hydrogen atoms during the *cis* bending vibration, $\nu_5$, cancel, and the vibration is not Raman active.

An **ellipsoid** is a three-dimensional body. For the most general type of ellipsoid, based on the *x*, *y* and *z* cartesian axes, cross-sections in each of the *xy*, *xz* and *yz* planes are elliptical.

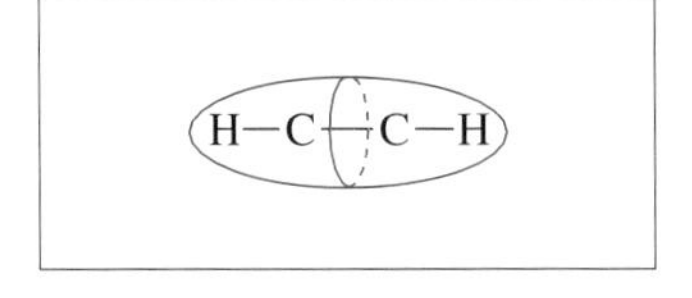


**Figure 10.4** A polarizability ellipsoid

Ethyne provides a good example of the **mutual exclusion rule**. This states that the infrared and Raman activity of vibrations in any molecule which has a centre of symmetry are mutually exclusive. Vibrations which are active in the infrared spectrum are not active in the Raman spectrum and *vice versa*.

Before we consider the rotational fine structure of only infrared vibrational spectra, it is necessary to understand the concept of a **transition moment**. When an infrared vibrational transition takes place, there is a change of dipole moment. Consider the simple example of HCl. Vibration causes an oscillating dipole moment which has a maximum value when the bond is at its longest and a minimum when it is at its shortest. The magnitude of the transition moment, which determines the intensity of the transition, is determined by the magnitude of the oscillating dipole moment. Its direction is clearly along the internuclear axis, and we say that the transition moment is **polarized** along that axis.

### Box 10.1 Vibrational Transition Moment

The transition moment, $\boldsymbol{R}_v$, for a transition between lower and upper states with vibrational wave functions $\psi''_v$ and $\psi'_v$, respectively, is given by:

$$\boldsymbol{R}_v = \int \psi'_v \, \boldsymbol{\mu} \psi''_v \, \mathrm{d}x \qquad (10.10)$$

where $x$ is the vibrational coordinate, $r - r_e$ for a diatomic molecule. The transition moment and the dipole moment, $\boldsymbol{\mu}$, are vector quantities. The transition intensity is proportional to $\boldsymbol{R}_v^2$.

In the case of a linear polyatomic molecule the oscillating dipole moment, and therefore the transition moment, may be polarized along the internuclear axis, to give a **parallel** band, or perpendicular to it, to give a **perpendicular** band. In the case of a parallel band, the rotational selection rule is the same as for a diatomic molecule:

$$\Delta J = \pm 1 \qquad (10.11)$$

resulting in a P-branch ($\Delta J = -1$) and an R-branch ($\Delta J = +1$), as illustrated in Figure 10.1.

In Figure 10.5, and also 10.7, the vertical scale represents the intensity of radiation transmitted by the sample. This way of showing the spectrum is quite common in infrared spectroscopy. Compared to a spectrum in which the degree of absorption is plotted on the vertical scale, these spectra appear to be "upside down".

Figure 10.5 shows a parallel band of the linear molecule $CO_2$. The vibration responsible is $\nu_3$, the asymmetric stretching vibration for which the transition moment is polarized along the internuclear axis. The band shows P- and R-branch structure similar to that of HCl in Figure 10.2. There is a typical rise and fall in the intensities along the P- and R-branches. There is a slight convergence in the R-branch and a divergence in the P-branch owing to the inequality of $B_0$ and $B_1$: $B_1$ is slightly less than $B_0$.

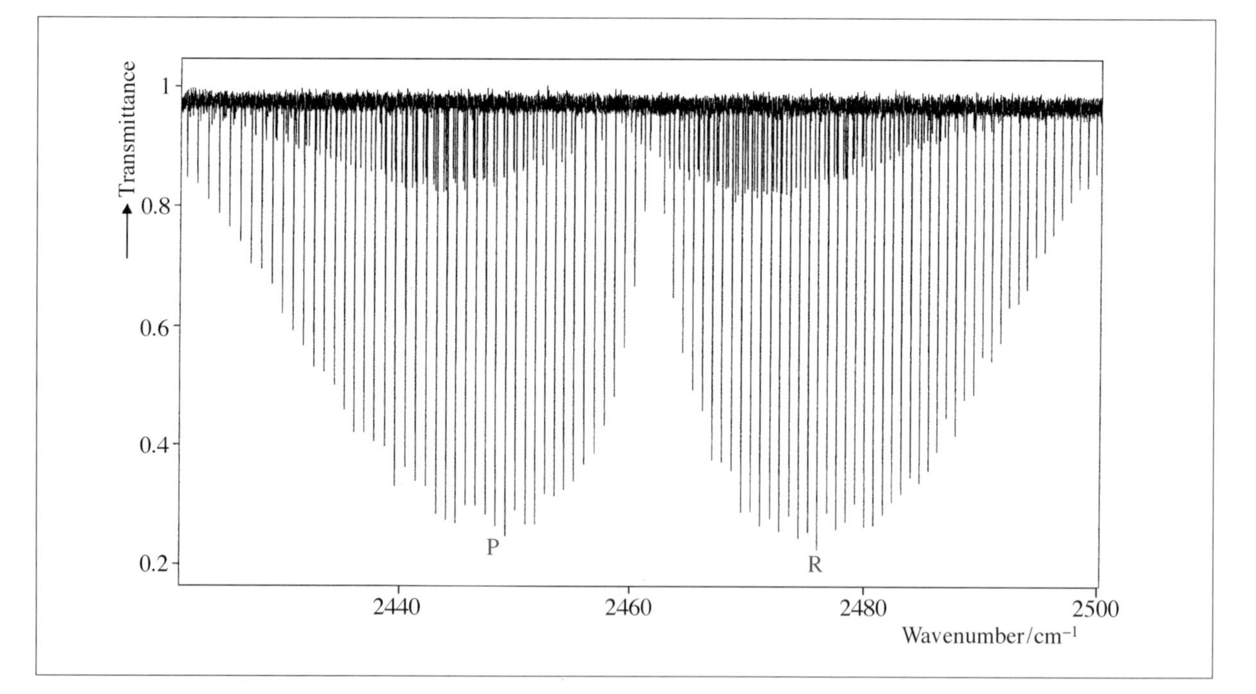


**Figure 10.5** Parallel band, involving the $v$ = 1–0 transition in the asymmetric stretching vibration, in the infrared spectrum of $CO_2$ (after G. Duxbury, *Infrared Vibration-Rotation Spectroscopy*, Wiley, New York, 2000, p. 176)

In Figure 10.5 there is a weak overlapping band, similar to the main band but displaced to low wavenumber. This is known as a **hot band** because its intensity increases with temperature. The band results from the population of a vibrational level involving one quantum of the bending vibration, $\nu_2$, and an allowed transition to a **combination level** involving one quantum each of $\nu_3$ and $\nu_2$. The appreciable population of the $v$ = 1 level of $\nu_2$ is caused by its having a relatively low wavenumber of 667 $cm^{-1}$.

### Worked Problem 10.2

**Q** Calculate the population, relative to the zero-point level, of the $v$ = 1 level of the $\nu_2$ vibration of $CO_2$ ($\omega$ = 667 $cm^{-1}$) at 293 K.

**A** Using equation (9.4), the relative population is given by:

$N_1/N_0 = 2 \times \exp(-hc \times 667\ cm^{-1}/kT)$
$= 2 \times \exp(-6.626 \times 10^{-34}\ J\ s \times 2.998 \times 10^{10}\ cm\ s^{-1} \times 667\ cm^{-1}/1.381 \times 10^{-23}\ J\ K^{-1} \times 293\ K)$
$= 0.0757$

This value, of about 7.6%, confirms the expectation that the hot band has appreciable intensity at 20 °C. Note that the bending vibration, $\nu_2$, is doubly degenerate, like $\nu_4$ and $\nu_5$ of ethyne, shown in Figure 6.6; hence the factor of two in the expression for $N_1/N_0$.

For a perpendicular band, in which the transition moment is perpendicular to the internuclear axis, the rotational selection rule is:

$$\Delta J = 0, \pm 1 \quad (10.12)$$

resulting in a P-branch ($\Delta J = -1$), a Q-branch ($\Delta J = 0$) and an R-branch ($\Delta J = +1$).

Figure 10.6 shows the stacks of rotational energy levels associated with the $v$ = 0 and 1 levels for a perpendicular band. A transition causing such a band is the $v$ = 1–0 transition involving the *cis* bending vibration, $\nu_5$, of the linear molecule ethyne, shown in Figure 6.6. The stack for the $v$ = 1 level differs from that for $v$ = 0 in that the $J$ = 1 level is missing, and the other rotational levels are split into doublets, the splitting increasing with $J$. Both these differences are a result of the double degeneracy of $\nu_5$. The double degeneracy confers on the vibration one quantum of vibrational angular momentum. Since $J$ represents the number of quanta of *total* angular momentum, its value can never be less than 1. Consequently, the first members of the R- and P-branches are R(0) and P(2), respectively.

There is a small splitting of the rotational levels of the $v$ = 1 state. This is due to a force, known as a **Coriolis force**, connecting the two components of $\nu_5$, which will not be considered further here. However, we shall see, in Section 11.2.3, that a similar splitting in an electronic spectrum is typically much larger.

The P- and R-branch wavenumbers can be used to obtain $B_0$ and $B_1$, as in equations (10.8) and (10.7), respectively.

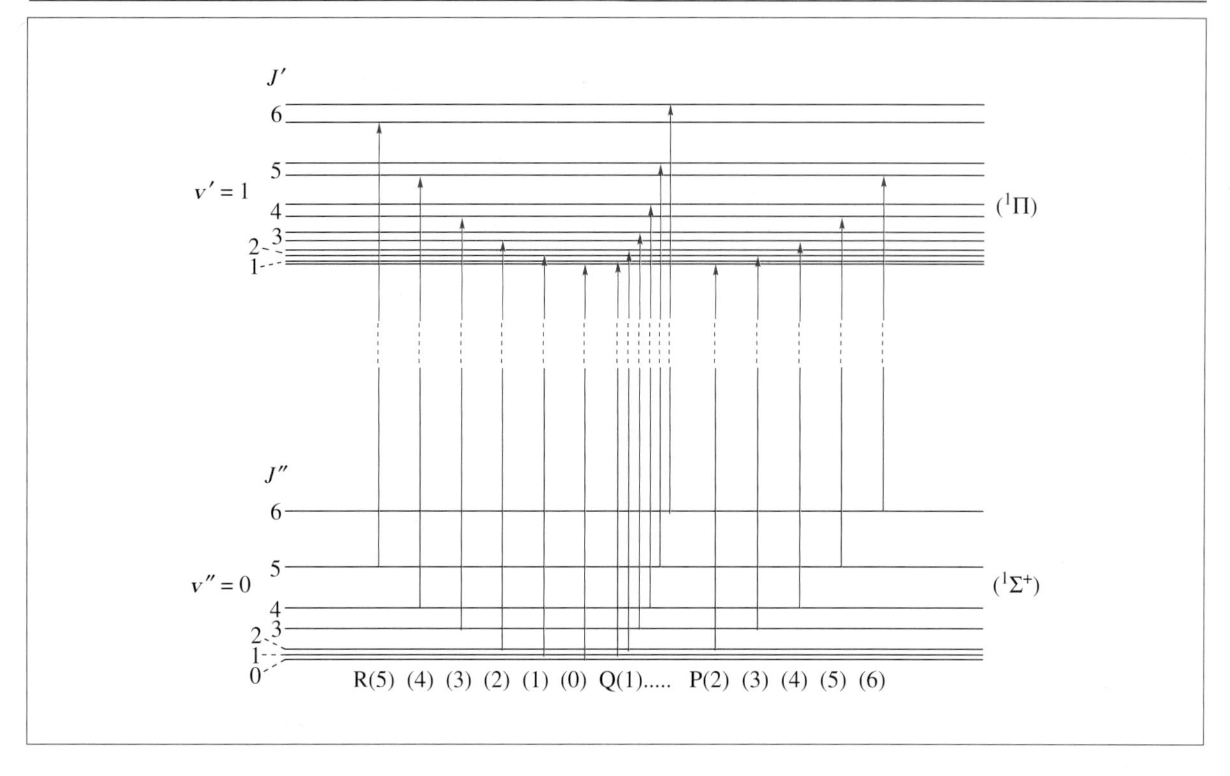


**Figure 10.6** Rotational energy levels and infrared-allowed transitions for a perpendicular band of a linear molecule (the $^1\Sigma^+$ and $^1\Pi$ symbols refer only to an electronic transition; see Chapter 11)

Figure 10.6 shows that using the P- and R-branch wavenumbers gives an effective value of $B_1$ for the *lower* components of the rotational levels of the $v = 1$ state. If the splitting of the rotational levels is large enough to be significant, the effective value of $B_1$ for the *upper* components can be obtained from the Q-branch. This will be discussed in more detail in Section 11.2.3 for the rotational fine structure of an electronic spectrum.

Figure 10.7 shows the $v$ = 1–0 transition involving the *trans* bending vibration $\nu_5$ of ethyne (see Figure 6.6). Ignoring the curvature in the baseline and the weak, overlapping hot bands, the main band can be seen to have a central Q-branch, with R- and P-branches to high and low wavenumber, respectively. The lines of the Q-branch are nearly coincident because of the near equality of the rotational constants $B_0$ and $B_1$. There is an observable intensity alternation in the R- and P-branches of 3:1 for $J$ odd:even. This is due to the two symmetrically equivalent hydrogen atoms having non-zero nuclear spin, with the nuclear spin quantum number $I = \frac{1}{2}$ (see Section 9.2.3).

## 10.5 Infrared and Raman Spectra of Symmetric Top Molecules

Symmetric top molecules have two equal principal moments of inertia (see Section 7.3.2). The rotational term values are given in equations (7.11) and (7.12) for a prolate and an oblate symmetric top, respectively, and are illustrated in Figure 7.5.

To be active in the infrared spectrum, the vibration must involve a change of dipole moment. Then, similar to the situation in a linear polyatomic molecule, the transition moment may be along the unique axis of the molecule (the *a*-axis in a prolate, the *c*-axis in an oblate

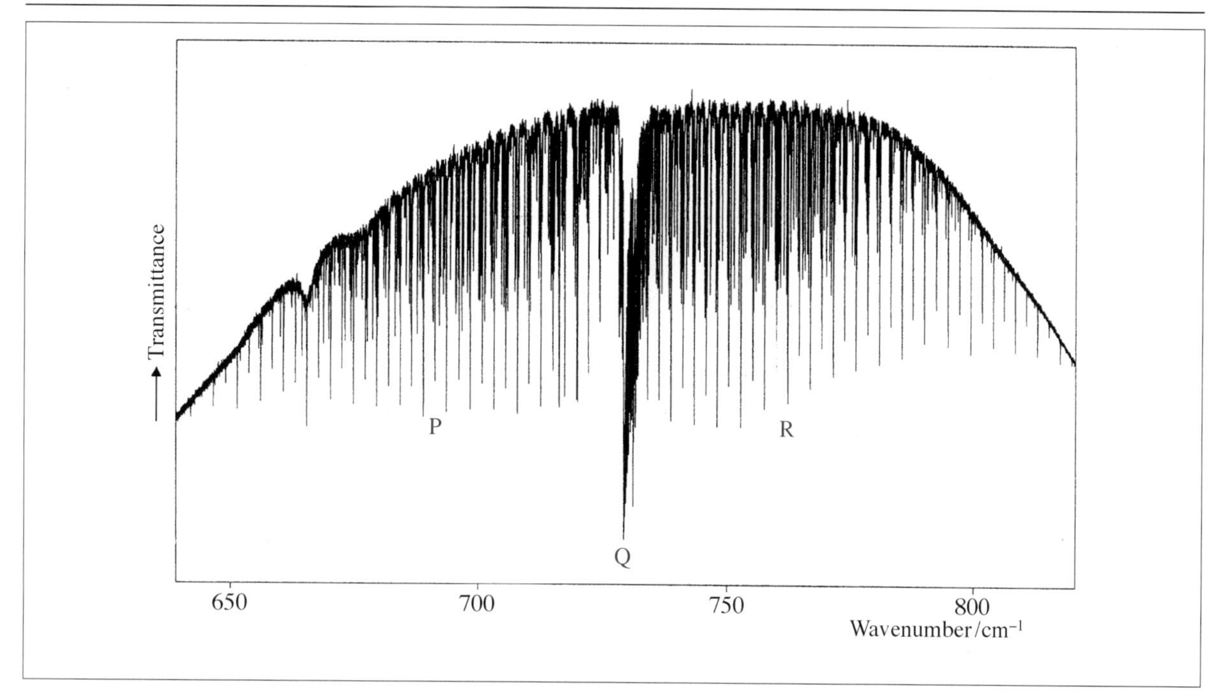


**Figure 10.7** Perpendicular band, involving the $v$ = 1–0 transition in the *trans* bending vibration, in the infrared spectrum of ethyne (acetylene) (after G. Duxbury, *Infrared Vibration-Rotation Spectroscopy*, Wiley, New York, 2000, p. 181)

symmetric top), to give a parallel band, or at 90° to it, to give a perpendicular band.

For a parallel band, the rotational selection rules are:

$$\Delta K = 0 \text{ and } \Delta J = \pm 1, \text{ for } K = 0$$
$$\Delta K = 0 \text{ and } \Delta J = 0, \pm 1, \text{ for } K \neq 0 \qquad (10.13)$$

The result is that there is a P-, Q- and R-branch for all values of $K$, except for $K = 0$ which has no Q-branch. For a typical vibrational transition the changes of rotational constants are very small. As a result, the P-, Q- and R-branches for all values of $K$ tend to lie on top of each other, making the band appear similar to that of a linear molecule.

Figure 10.8 shows a parallel band, centred at about 1048 cm$^{-1}$, of the prolate symmetric top fluoromethane, shown in Figure 7.3(a). The vibration responsible for this band is $\nu_3$, the C–F stretching vibration, for which the dipole moment change is clearly along the $a$-axis.

The spectrum in Figure 10.8 also shows a perpendicular band centred at about 1182 cm$^{-1}$ and due to $\nu_6$, a vibration involving a rocking motion of the hydrogen atoms. The dipole moment change is in a direction at 90° to the $a$-axis.

The selection rules for a perpendicular band are:

$$\Delta K = \pm 1 \quad \text{and} \quad \Delta J = 0, \pm 1 \qquad (10.14)$$

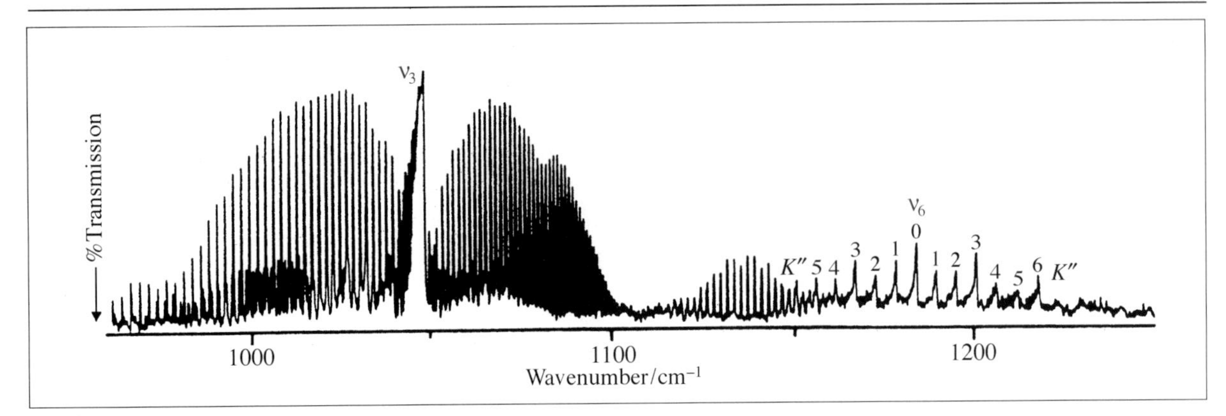


**Figure 10.8** Parallel ($\nu_3$) and perpendicular ($\nu_6$) bands in the infrared spectrum of fluoromethane (after J. M. Hollas, *High Resolution Spectroscopy*, 2nd edn., Wiley, New York, 1998, p. 230)

The Q-branches of $\nu_6$ in Figure 10.8 show relative intensities of 4, 2, 2, 4, ... for $K''$ = 0, 1, 2, 3, 4, ... . This is a consequence of the nuclear spin ($I = \frac{1}{2}$) of the three symmetrically equivalent hydrogen atoms.

The effect of the $\Delta K = \pm 1$ selection rule, compared to $\Delta K = 0$ for a parallel band, is to spread out the P-, Q- and R-branches with different values of $K$. Because the rotational constants for different vibrational states tend to be very similar, the Q-branches consist of nearly coincident lines and therefore dominate the spectrum. This is the case for $\nu_6$ in Figure 10.8, where the Q-branches are labelled with the corresponding value of $K''$ in the zero-point level.

## Worked Problem 10.3

**Q** Show that the separation of Q-branches in the infrared vibrational spectrum of a prolate symmetric top is approximately $2(A - B)$. What is the corresponding separation for an oblate symmetric top?

**A** We use equation (7.11) for the rotational term values of a prolate symmetric top. Then we assume that (a) we can neglect any centrifugal distortion terms in that equation, and (b) there is no change of rotational constants from the lower to the upper vibrational state. Then, applying the $\Delta K = \pm 1$, $\Delta J = 0$ (for Q-branches) selection rules, gives:

$$F(J,K \pm 1) - F(J,K) = BJ(J+1) + (A-B)(K \pm 1)^2 - BJ(J+1) - (A-B)K^2 = \pm 2K(A-B) + 2(A-B) \quad (10.15)$$

When $K$ changes by one, this difference, which is the separation of adjacent Q-branches, changes by $2(A - B)$. For an oblate symmetric top, the rotational term values are given by equation (7.12), in which the rotational constant $A$ has been replaced by $C$. Therefore, the Q-branch separations are approximately $2(C - B)$.

Raman spectra of symmetric top molecules in the gas phase are difficult to obtain with sufficiently high resolution to show rotational fine structure, and we shall not consider them further. However, liquid phase Raman spectra with lower resolution and showing just the vibrational bands are much more readily observed. Such spectra are of particular importance when used as an analytical tool for identifying either individual molecules or groups within a molecule. These important aspects of vibrational spectroscopy will be discussed further in Section 10.6, but the vibrational spectrum of benzene (see Figure 7.3b), which is a symmetric top, is shown here in Figure 10.9. This figure shows both the Raman and infrared spectra of liquid benzene, in which any rotational structure is removed owing to intermolecular interactions.

**Figure 10.9** Infrared (upper) and Raman (lower) spectra of liquid benzene (after B. Schrader, *Raman/Infrared Atlas of Organic Compounds*, VCH, Weinheim, 1989, p. F1-01)

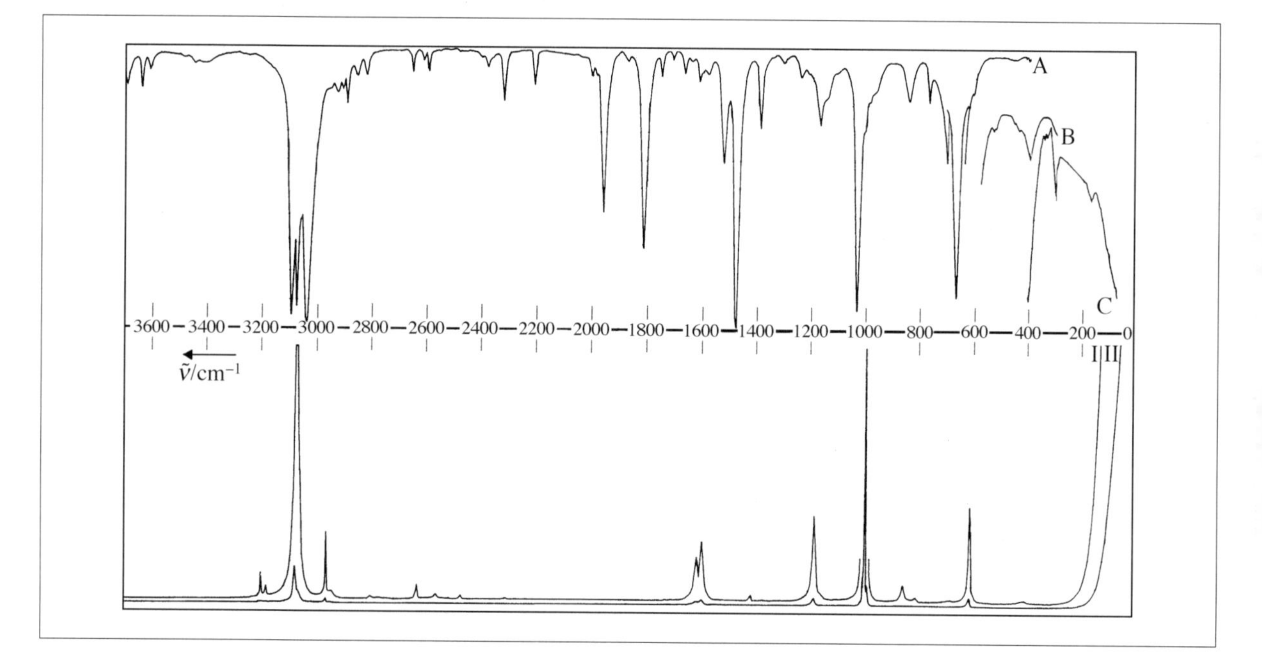


Of particular interest in the case of benzene is the importance of the mutual exclusion rule, introduced in Section 10.4. A consequence of the molecule having a centre of symmetry is that vibrations which are active in the infrared spectrum are inactive in the Raman spectrum, and *vice versa*.

It may appear that the vibration at about 3060 $cm^{-1}$ in the Raman spectrum of benzene is also observed in the infrared spectrum. This is the region of C–H stretching vibrations. There are six of these in benzene, all of them in the region 3000–3100 $cm^{-1}$. It is accidental that one which is active in the infrared

## 10.6 Infrared and Raman Spectra of Asymmetric Top Molecules

For small asymmetric top molecules, such as $H_2O$, the rotational fine structure of vibrational bands in the infrared spectrum is readily resolved. However, in line with the policy adopted elsewhere in this book, the rotational energy levels and transitions of asymmetric tops will not be considered further. Suffice it to say that the rotational constants and geometrical structure of many small asymmetric top molecules have been obtained using high-resolution, gas phase, infrared vibrational spectroscopy. High-resolution Raman vibrational spectra are much less often studied.

The vast majority of molecules are relatively large and heavy asymmetric tops. Consequently, the rotational constants *A*, *B* and *C* (see equations 7.9 and 7.10) are small, resulting in very congested rotational fine structure of vibrational bands. In order to observe the rotational structure, the molecule must be in the gas phase with a sufficiently high vapour pressure, which is not possible with many solid materials. Even if the spectrum can be obtained, the rotational transitions in a high-resolution, gas phase, infrared spectrum may be so congested that only a **contour**, or envelope, of the underlying, unresolved rotational structure is observed.

Figure 10.10 shows part of the infrared absorption spectrum of fluorobenzene in the gas phase. The molecule is an asymmetric top with such small rotational constants that only rotational contours can be observed for each vibrational band. Although the molecule is an asymmetric top, it has unique axes, defined by the symmetry of the molecule, which also correspond to the unique inertial axes, *a*, *b* and *c*, as shown in Figure 10.10.

For fluorobenzene, $A_0$ = 0.18892 cm$^{-1}$, $B_0$ = 0.08575 cm$^{-1}$ and $C_0$ = 0.05897 cm$^{-1}$, in the zero-point level.

Symmetry plays a very important part in the theory of all types of spectra, particularly concerning selection rules, but is too large a subject to be discussed in detail in this book. Books which cover symmetry in detail are listed in "Further Reading" at the end of this book. However, in the case of fluorobenzene, it is fairly obvious that, although the inertial axes are confined by the symmetry of the molecule, they are not similarly confined in a molecule with no symmetry at all, such as the tetrahedral molecule CHBrClF.

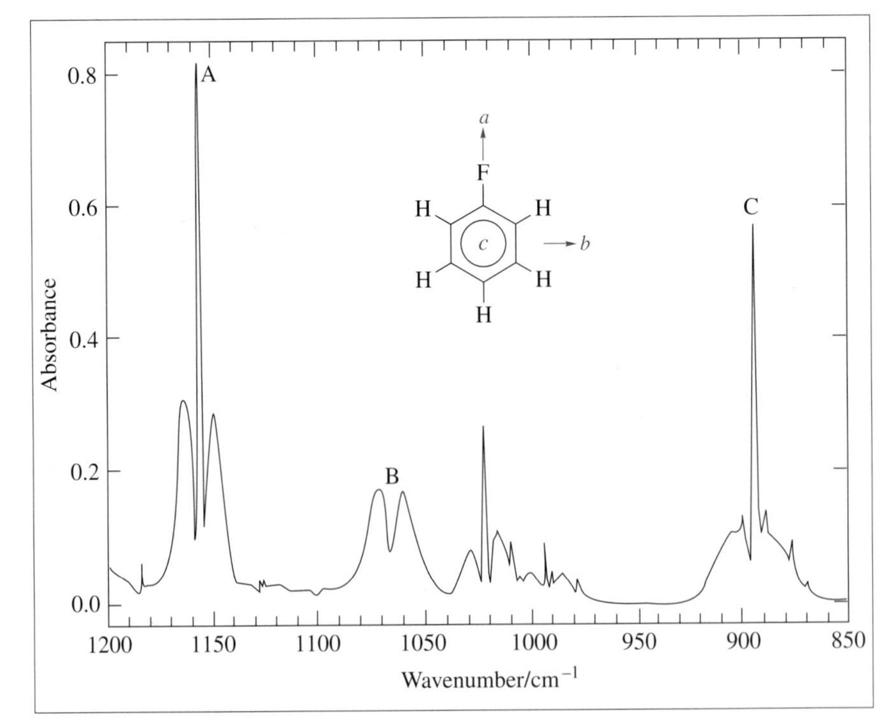


**Figure 10.10** Mid-infrared spectrum of fluorobenzene in the gas phase

Therefore, the transition moment for any vibrational transition is polarized along one of these axes: a type A, B or C band is polarized along the *a*, *b* or *c* axis, respectively. The rotational selection rules are different for each band type. Because rotational constants vary very little from the zero-point level to excited vibrational levels, all three band types exhibit characteristic rotational contours, irrespective of the molecule or the vibrational transition. A type A band typically shows an intense, sharp central spike with quite intense, broad wings on either side. The band at about 1156 $cm^{-1}$ in Figure 10.10 is a typical type A band. The transition moment is polarized along the axis containing the C–F bond. The band at about 1067 $cm^{-1}$ is a typical type B band, showing a central, deep minimum and broad wings resembling those of a type A band. The transition moment is along the *b*-axis. A type C band shows an intense, sharp central spike, rather like a type A band. It also shows broad wings on either side but of lower intensity, relative to the central spike, than in a type A band. The band at about 893 $cm^{-1}$ is a typical type C band; the transition moment lies along the *c*-axis, perpendicular to the plane of the molecule.

### Worked Problem 10.4

**Q** What other band types can you identify in the spectrum in Figure 10.10?

**A** The band at about 1022 $cm^{-1}$ is clearly a type A band, although it suffers some interference from a neighbouring band on the low wavenumber side. The band at about 993 $cm^{-1}$ is not so straightforward. It is either a type A or type C band but the intensity of the wings, relative to the central spike, is not easy to judge. There are numerous weak bands in the spectrum characterized by a sharp spike. It is not possible to assign them unambiguously as type A or C bands. However, the more intense central spike of a type C band makes it rather more likely that a weak type C band would be detected. Weak type B bands are not easily observed.

One of the aims of obtaining a gas phase spectrum, showing characteristic rotational contours, is to identify the direction of the transition moment and use this to aid the attribution of a band as being due to a particular vibrational motion. For example, the transition moment corresponding to the type A band of fluorobenzene, at 1156 $cm^{-1}$, lies along the *a*-axis (see Figure 10.10). Therefore the change of dipole moment during this vibration is along that axis. This is consistent with the expectation of a C–F stretching vibration (a group vibration: see Section 6.3) having a wavenumber of this magnitude. The identification of the type

In fact, the 1156 $cm^{-1}$ vibration in fluorobenzene does not involve a pure C–F stretching motion but also includes some vibrational motion of the benzene ring. This has been shown by deuteration of the hydrogen atoms. The resulting significant reduction of the vibration wavenumber is due to the involvement of the heavier deuterium atoms.

A contour confirms the assignment of the C–F stretching vibration. The high intensity of the band is a result of a relatively large change of dipole moment accompanying the stretching and contraction of the C–F bond.

Figure 10.11 shows the infrared and Raman spectra of cyanobenzene (benzonitrile). Like the spectra of benzene in Figure 10.9, these spectra are of a pure liquid for which only a small amount of sample is required. The path lengths used for the infrared spectra A, B and C of benzene are 40, 500 and 1500 μm, respectively, and, for the infrared spectra A and B of benzonitrile, 25 and 1000 μm, respectively.

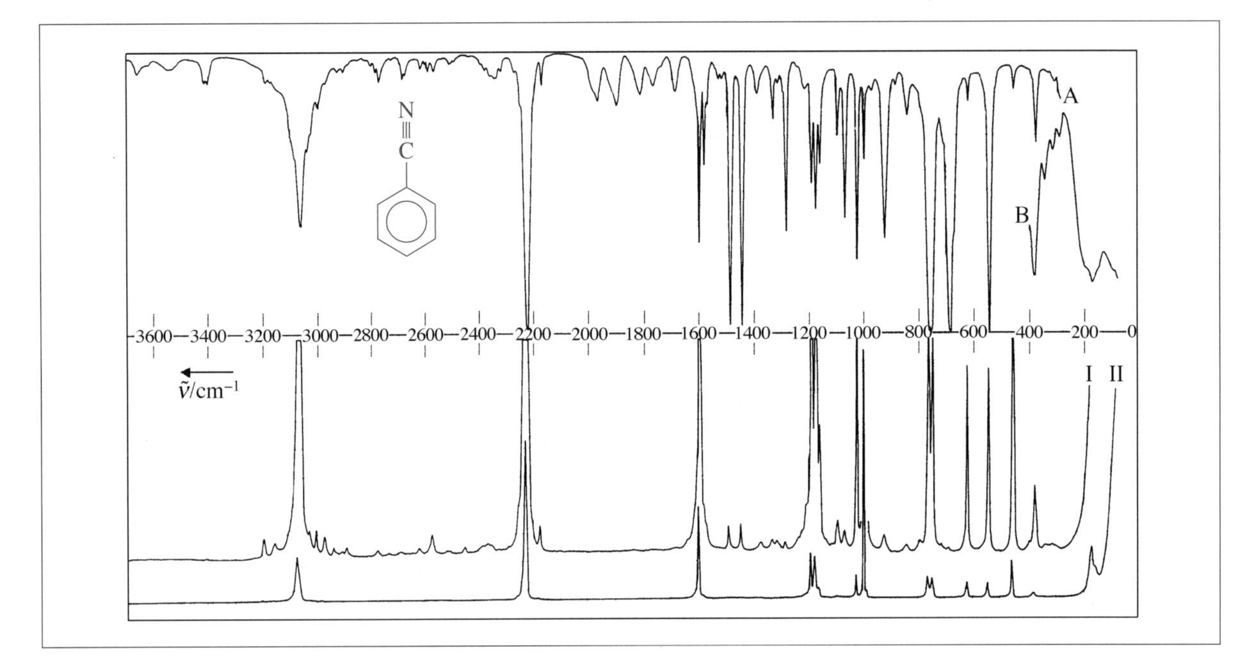


**Figure 10.11** Infrared (upper) and Raman (lower) spectra of liquid cyanobenzene (benzonitrile) (after B. Schrader *Raman/Infrared Atlas of Organic Compounds*, VCH, Weinheim, 1989, p. F1-06)

The spectra in Figures 10.9 and 10.11 illustrate some important features of infrared and Raman spectra of molecules in a condensed phase. Here, the spectra are of the pure liquid but are also observed, commonly, either in solution in a transparent solvent or in the form of a fine powder, mixed with potassium bromide, and compressed into a **KBr disk**. Although the infrared and Raman spectra of benzene show the mutual exclusion characteristic of a molecule with a centre of symmetry (see Section 10.5), the spectra of benzonitrile, which does not have a centre of symmetry, show many bands which appear in both spectra. These spectra show how molecules of this size have very many characteristic bands, particularly molecules with no centre of symmetry.

Group vibrations are particularly important as an aid to identification of a molecule. For example, the success of the incorporation of a C=O group into a molecule by a chemical reaction may be judged by the appearance of a very strong band in the infrared spectrum in the region 1650–1830 $cm^{-1}$, due to stretching of the C=O bond. Although this is quite a large range, the stretching motion involves such a large change of dipole moment that the band often dominates the infrared spectrum to such an extent that there can be little doubt that the molecule contains a C=O group.

Although many stretching vibrations show typical wavenumbers, this applies also to some bending vibrations. Table 10.1 shows typical wavenumbers of some stretching and bending vibrations.

The use of group vibrations in analytical infrared and Raman spectroscopy is most common in organic molecules, or in inorganic complexes containing organic ligands. Nevertheless, it is worth noting that C=O stretching vibrations in inorganic metallic carbonyls may have a wavenumber in excess of 2000 $cm^{-1}$.

**Table 10.1** Some typical bond stretching and bending vibration wavenumbers

| *Group* | *Wavenumber/$cm^{-1}$* | *Group* | *Wavenumber/$cm^{-1}$* |
|---|---|---|---|
| >C–H stretch | 2960 | >C–F stretch | 1100 |
| =C<H stretch | 3020[a] | –O–H stretch | 3600[b] |
| ≡C–H stretch | 3300 | >N–H stretch | 3350 |
| –C–C– stretch | 900 | –C<H₃ bend | 1000 |
| >C=C< stretch | 1650 | =C<H₂ bend | 1100 |
| –C≡C– stretch | 2050 | ≡C–H bend | 700 |
| >C=O stretch | 1700 | >C<H₂ bend | 1450 |
| –C≡N stretch | 2100 | C≡C–C bend | 300 |

[a] Except in the CHO group when a typical wavenumber is 2800 $cm^{-1}$.
[b] This is considerably reduced in a hydrogen-bonding solvent.

The wavenumbers of the three types of C–H stretching vibrations show how these differ according to the hybridization of the 2s and 2p atomic orbitals on the carbon atom. When the carbon atom of the C–H group takes part in three single, one double or one triple bond, the hybridization is assumed to be $sp^3$, $sp^2$ or sp, respectively. A bond involving a hybrid orbital is stronger the more s-character it has and consequently has a larger stretching wavenumber. This is consistent with the C–H group having the largest wavenumber when including a carbon atom involved in a triple bond, and the smallest when it is involved in three other single bonds. Similarly, the stretching wavenumber of a carbon–carbon bond increases with the multiplicity of the bond, being largest for a triple bond.

In general, angle bending vibrations are not such useful group vibrations as stretching vibrations. Table 10.1 shows five bending group vibrations involving carbon and hydrogen atoms.

Assignment of bands in an infrared or Raman spectrum to particular vibrational motions in the molecule is more readily achieved at wavenumbers above about 1500 $cm^{-1}$. Below this, the region 1500–600 $cm^{-1}$ is referred to as the **fingerprint region**. This is a region where many vibrational motions are not localized, even approximately, in a particular group in the molecule. We can regard the hypothetical localized motions as being **coupled** so that the delocalized vibrational motion involves more atoms in the molecule. The fingerprint region is so-called because the wavenumbers of these delocalized vibrations tend to be characteristic of a particular molecule. Consequently, the fingerprint region is very important in the unique identification of a molecule rather than of individual groups within a molecule.

The spectra of benzene and benzonitrile, in Figures 10.9 and 10.11, show bands due to C–H stretching vibrations in the region of 3050 $cm^{-1}$. These are characteristic of a carbon atom attached to a carbon atom with $sp^2$ hybridization. The fingerprint region, below about 1500 $cm^{-1}$, is more complex, particularly for benzonitrile, because of its much lower symmetry. The infrared and Raman spectra of benzonitrile show, very strongly, the C≡N stretching group wavenumber at 2226 $cm^{-1}$, and the infrared spectrum of fluorobenzene in Figure 10.10 shows a C–F stretching group wavenumber of 1165 $cm^{-1}$.

In Section 2.3 the clear distinction between frequency and wavenumber was made. From this it follows that we should speak of "group wavenumbers". However, this distinction is often ignored so that we shall often encounter "group frequencies", even though they are given with units of wavenumber.

## Summary of Key Points

**1.** *Infrared vibrational spectra of diatomic molecules*
Vibrational selection rule. Selection rule for rotational fine structure. Requirement that the molecule has a permanent dipole moment. Determination of the rotational constant, *B*, and internuclear distance, in both combining vibrational states.

**2.** *Raman spectra of diatomic molecules*
Vibrational selection rule. Selection rule for rotational fine structure. All diatomic molecules show a Raman spectrum.

**3.** *Infrared and Raman spectra of linear polyatomic molecules*
Normal vibrations of a linear molecule. Infrared and Raman activity are mutually exclusive for a symmetrical linear polyatomic molecule. An infrared active vibration involves a change of dipole moment while a Raman-active vibration involves a change of polarizability.

**4.** *Concept of a vibrational transition moment*
Rotational selection rules for an infrared vibrational transition

depend on whether the transition moment is parallel or perpendicular to the internuclear axis of a linear polyatomic molecule.

5. *Hot bands and combination levels*
Hot vibrational bands appear when a molecule has one or more vibrations of low wavenumber. A combination level involves quanta (one or more) of more than one vibration.

6. *Infrared spectra of symmetric top molecules*
Vibration must involve a change of dipole moment. Selection rules for parallel and perpendicular bands.

7. *Infrared spectra of asymmetric top molecules in the gas phase*
In large asymmetric top molecules, rotational fine structure cannot be resolved. Only a characteristic rotational contour is observed. In molecules which have some symmetry, typical type A, B or C contours may be observed when the transition moment is along the *a*, *b* or *c* inertial axis.

8. *Infrared and Raman spectra of large asymmetric top molecules in a condensed phase*
Use as analytical tools. Concept of a group vibration and use in identifying a group within a molecule. Unique pattern of vibrational bands in either the infrared or Raman spectrum used to identify a particular molecule.

## Problems

**10.1.** In the vibrational Raman spectrum of CO, the O(10) and S(10) rotational transitions are separated by 160.2 $cm^{-1}$. Use this information to calculate an approximate bond length for CO. Which approximations are involved? [$m(^{12}C) = 12.00$ u, $m(^{16}O) = 15.99$ u]

**10.2.** How many vibrations does the $CO_2$ molecule have? Describe the motions involved and comment on their infrared and Raman activity.

**10.3.** What typical group stretching vibrations would you expect in the propyne ($CH_3$–C≡C–H) molecule?

# 11
# Electronic Spectroscopy

## Aims

In Chapter 5, electronic orbitals and states of diatomic and polyatomic molecules were discussed. By the end of the present chapter you should be able to understand:

- The selection rules for electronic spectra of diatomic molecules
- The Franck–Condon principle and its application to vibrational coarse structure of electronic spectra of diatomic molecules
- Vibrational progressions and sequences
- Rotational fine structure in $\Sigma$–$\Sigma$ and $\Pi$–$\Sigma$ types of electronic transitions in diatomic molecules
- How bond lengths in ground and excited electronic states of diatomic molecules may be obtained
- The general features of the vibrational coarse structure and the rotational fine structure of electronic spectra of polyatomic molecules
- The uses of electronic spectra of molecules in solution for identifying chromophores and determining concentrations

## 11.1 Introduction

In Chapter 5, the orbitals and electronic states of diatomic and polyatomic molecules were discussed. In the present chapter, the investigation of these states by the observation of transitions between them, by electronic spectroscopy, will be described. As was the case for rotational and vibrational spectroscopy in Chapters 9 and 10, discussion of electronic spectroscopy will be mainly concerned with absorption spectra.

Associated with each electronic state are vibrational and rotational energy levels, as illustrated for a diatomic molecule in Figure 6.9. The rotational fine structure of electronic spectra has much in common with that of vibrational spectra, but the vibrational coarse structure is more complex.

Electronic spectroscopy is employed as an analytical tool, particularly for large molecules. As in analytical vibrational spectroscopy, such electronic spectra are usually obtained in the liquid phase. However, the unique character of a vibrational spectrum, allowing the identification of a molecule, or a group within a molecule, is not so apparent in an electronic spectrum. On the other hand, the application of the Beer–Lambert law, discussed in Section 8.1, to an electronic spectrum allows the measurement of concentration of a molecule whose molar absorption coefficient (see equation 8.2) is known.

## 11.2 Electronic Spectra of Diatomic Molecules

### 11.2.1 Electronic States and Selection Rules

In Sections 5.1.2 and 5.2.2 the symbolism for ground and excited electronic states of homonuclear and heteronuclear diatomic molecules was developed. There are two quantum numbers which will concern us and which feature in these symbols: $\Lambda$ refers to the component of the orbital angular momentum of the electrons along the internuclear axis, and $S$ is the total electron spin quantum number. The main symbol for an electronic state is $\Sigma$, $\Pi$, $\Delta$, $\Phi$, ... and indicates the value of $\Lambda$, as shown in equation (5.10). The pre-superscript on the main symbol is the multiplicity, $2S + 1$, of the state.

#### Box 11.1 Labelling of States Split by Spin–Orbit Coupling

The value of $\Sigma$, defined in equation (5.11), refers to the component of the electron spin angular momentum along the internuclear axis, and may be included in the state symbol only for states with $\Lambda > 0$ and multiplicity $(2S + 1) > 1$. This is the case, for example, in a $^3\Pi$ state, where $\Lambda = 1$ and $S = 1$. Then, $\Sigma = 1$, 0 and $-1$. If the three components of the $^3\Pi$ state are split by spin–orbit coupling, which is appreciable if the molecule contains at least one heavy atom, then the value of $\Lambda + \Sigma$ is indicated by a post-subscript, as in the $^3\Pi_2$, $^3\Pi_1$ and $^3\Pi_0$ components of the $^3\Pi$ state. Splitting of such components, and the necessity of indicating the value of $\Lambda + \Sigma$, will not be encountered subsequently in this book.

As discussed in Section 5.1.2, the electronic wave function for a $\Sigma$ state may be symmetric or antisymmetric to reflection through any plane containing the internuclear axis: such states are labelled $\Sigma^+$ or $\Sigma^-$, respectively. For homonuclear diatomic molecules the electronic wave function is either symmetric or antisymmetric to reflection through the centre of the molecule. This is indicated (see Section 5.1.1) by a post-subscript "g" or "u", respectively, as in, for example a ${}^3\Sigma_g^+$ and a ${}^2\Pi_u$ state.

All molecules have many excited electronic states, in addition to the ground state. For diatomic molecules, we shall be concerned only with bound states. These have a deep minimum in the potential energy curve, like those in Figure 6.9, unlike unbound, repulsive states such as that in Figure 6.10.

The selection rule in the quantum number $\Lambda$ is:

$$\Delta\Lambda = 0, \pm 1 \tag{11.1}$$

Consistent with the convention adopted elsewhere, an electronic transition between, for example, a lower $\Sigma$ state and an upper $\Pi$ state is indicated by $\Pi$–$\Sigma$. The selection rule applies also if the states are reversed; for example, a $\Sigma$–$\Pi$ as well as a $\Pi$–$\Sigma$ transition is allowed.

For example, $\Sigma$–$\Sigma$, $\Pi$–$\Pi$ and $\Pi$–$\Sigma$ transitions are allowed, but $\Delta$–$\Sigma$ and $\Phi$–$\Pi$ transitions are forbidden.

Concerning electron spin, the selection rule is:

$$\Delta S = 0 \tag{11.2}$$

but this rule breaks down when one or more heavy atom is present. For example, a ${}^3\Pi_u$–${}^1\Sigma_g^+$ transition is observed only very weakly in $O_2$, but strongly in $I_2$.

The symmetry selection rule:

$$+ \leftrightarrow +,\ - \leftrightarrow -\ \text{and}\ + \not\leftrightarrow - \tag{11.3}$$

where $\leftrightarrow$ and $\not\leftrightarrow$ signify allowed and forbidden transitions, respectively, applies to $\Sigma$–$\Sigma$ transitions: only $\Sigma^+$–$\Sigma^+$ and $\Sigma^-$–$\Sigma^-$ transitions are allowed.

For homonuclear diatomic molecules, there is an additional selection rule:

$$\text{g} \leftrightarrow \text{u},\ \text{g} \not\leftrightarrow \text{g}\ \text{and}\ \text{u} \not\leftrightarrow \text{u} \tag{11.4}$$

For example, ${}^1\Pi_g$–${}^1\Sigma_g^+$ and ${}^1\Sigma_u^-$–${}^1\Sigma_u^-$ transitions are forbidden but ${}^3\Pi_u$–${}^3\Pi_g$ and ${}^1\Delta_g$–${}^1\Pi_u$ transitions are allowed.

In Section 11.2.3 we shall see that, in cases where the lower state of an electronic transition is the ground state, the rotational constant, $B$, and consequently the bond length, $r$, can be obtained. However, provided the electronic transition is allowed, this method has an important advantage over the use of rotational or vibrational infrared spectroscopy in that it can be used for both homonuclear and heteronuclear diatomic molecules.

**Worked Problem 11.1**

**Q** Deduce all the types of electron transitions which are allowed for which the lower state is $^3\Delta_u$ in (a) $H_2$ and (b) $I_2$.

**A** (a) $H_2$ is a very light molecule, and the $\Delta S = 0$ selection rule holds rigidly. Therefore, the upper state of a transition must be a triplet state also. Then, applying the selection rules of equations (11.1), (11.2) and (11.4), the allowed transitions are $^3\Phi_g$–$^3\Delta_u$ and $^3\Pi_g$–$^3\Delta_u$.
(b) $I_2$ is a heavy molecule. The $\Delta S = 0$ selection rule breaks down, so that, in addition, transitions with $\Delta S = \pm 1$ (but not $\pm 2, \pm 3, \ldots$) may be observed. Therefore the upper electronic state could have $S = 0$ or 2, resulting in a singlet or quintet state, respectively. The selection rules of equations (11.1) and (11.4) remain valid. The allowed transitions involving $^3\Delta_u$ as a lower state are $^5\Phi_g, {}^3\Phi_g, {}^1\Phi_g$–$^3\Delta_u$ and $^5\Pi_g, {}^3\Pi_g, {}^1\Pi_g$–$^3\Delta_u$.

### 11.2.2 Vibrational Coarse Structure

As indicated in Figure 6.9, each bound electronic state, whether it is the ground state or an excited state, has its own potential function characterized by the equilibrium internuclear distance, $r_e$, and the dissociation energy, $D_0$ (or $D_e$). Each function is also characterized by a set of vibrational energy levels, converging smoothly towards the dissociation limit, with term values given by:

$$G(v) = \omega_e(v + \tfrac{1}{2}) - x_e(v + \tfrac{1}{2})^2 + y_e(v + \tfrac{1}{2})^3 + \ldots \qquad (11.5)$$

This is the same as equation (6.14), which we applied only to the ground electronic state. When applied to other electronic states, the values of $\omega_e$, $x_e$, $y_e$, ... apply uniquely to each state.

The vibrational transitions which accompany all allowed electronic transitions are referred to as **vibronic transitions**. If we are concerned with an absorption process, most of the molecules will be, initially, in the $v'' = 0$ state of the ground electronic state.

The selection rule governing these vibronic transitions is completely unrestrictive:

$$\Delta v = 0, \pm 1, \pm 2, \pm 3, \ldots \qquad (11.6)$$

In an absorption spectrum, the vibronic transitions from $v'' = 0$ form a

Consistent with the convention used previously, $v''$ indicates the vibrational quantum number of the lower state (often the ground state) of an electronic transition and $v'$ that of the upper state.

**progression**. Although the selection rule allows transitions with all values of $v'$, it is the intensity distribution along the progression that determines which transitions are sufficiently intense to be observed. This distribution is governed by the **Franck–Condon principle**

Initially, it was Franck who considered the intensities of vibronic transitions using only classical, as opposed to the quantum mechanical, considerations. Classically, there is no zero-point energy level, and the extremities of each vibrational energy level, where it meets the potential energy curve, are classical turning points of the vibration. At these points the nuclei are momentarily stationary, at a position of either maximum extension or compression of the bond. Since electrons are much lighter than nuclei, an electronic transition takes place much more rapidly than vibrational motion. Consequently, the vibronic transition with the greatest probability is that which starts from the bottom of the ground state potential energy curve, at which point the nuclei are stationary, and goes to a point on the excited state potential energy curve where the nuclei are also stationary, *i.e.* the bond length $r$ remains constant during the transition. Such a transition is said to be a **vertical transition** because, on a figure showing the ground and excited state potential energy curves, the transition is represented by a vertical line.

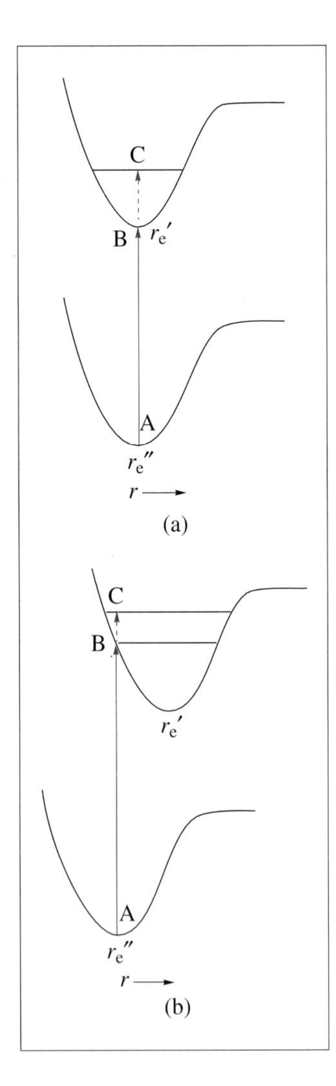


**Figure 11.1** Transitions B–A are the most probable (classical) vibronic transitions when (a) $r_e' \approx r_e''$ and (b) $r_e' > r_e''$

Consistent with the convention used throughout spectroscopy, the transition indicated by B–A implies that B is the *upper* state and A is the *lower* state. In general, a vibronic transition is indicated by $v'$–$v''$.

Figure 11.1 shows two examples of vertical transitions. In Figure 11.1(a) the equilibrium internuclear distance, $r_e$, is the same in both electronic states. As a result, the most probable vertical transition, indicated by B–A, is to the minimum of the excited state potential energy curve. Any other possible vertical transition, such as C–A, would be to a point in the vibrational level C which is far away from the classical turning points and at which the nuclei would be moving at maximum speed. As a result, this transition is highly improbable.

The situation illustrated in Figure 11.1(b) is one in which $r_e' > r_e''$. Then, the most probable transition, B–A, is to the turning point of an excited vibrational level. Vertical transitions, such as C–A, to vibrational levels above and below B are less probable, and therefore less intense, because the nuclei are not stationary.

Condon considered the consequences of quantum mechanical treatment of vibrational motion. As shown in Figure 11.2, the lowest vibrational energy level is not at the minimum of the potential energy curve: $v = 0$ corresponds to the zero-point energy. In addition, the molecule no longer spends most of the time exactly at the classical turning point of a vibration, but where the vibrational wave function has its maximum value. This is at the centre of the zero-point level, as shown in the figure. For the $v = 1$, 2 and 3 vibrational levels it is quite far from the classical turning points, but as $v$ increases the maximum in the wave function moves closer to the classical turning point.

Figure 11.2 shows the quantum mechanical equivalents of the situa-

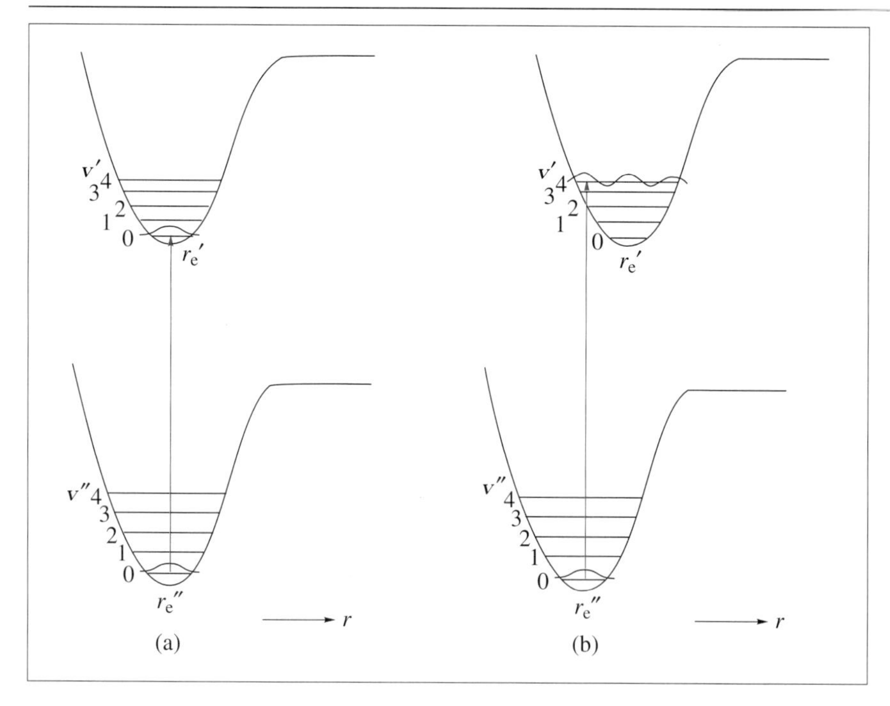


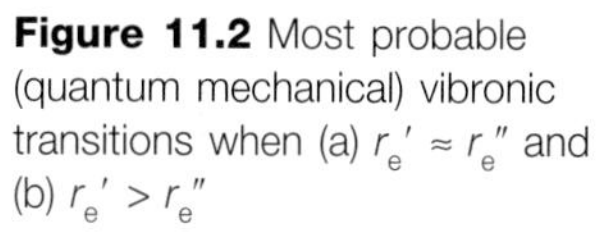
**Figure 11.2** Most probable (quantum mechanical) vibronic transitions when (a) $r_e' \approx r_e''$ and (b) $r_e' > r_e''$

tions in Figure 11.1. In Figure 11.2(a) the most probable transition is from the mid-point of the $v'' = 0$ level of the ground electronic state to the mid-point of the $v' = 0$ level of the excited electronic state, but in Figure 11.2(b) it is to a point close to the classical turning point of the $v' = 4$ level.

The Franck–Condon principle states that the most probable vibronic transition is a vertical transition between positions on the vibrational levels of the upper and lower electronic state at which the vibrational wave functions have maximum values.

Because of the smeared-out nature of the vibrational wave functions, vibronic transitions such as the 0–0 and 4–0 in Figure 11.2(a) and 11.2(b), respectively, are not the only transitions with non-zero intensity. For the case illustrated in Figure 11.2(a) there is a short progression in which the intensity decreases rapidly with increasing $v'$, as shown in Figure 11.3(a). When there is an appreciable increase of $r$ in the excited state, as in Figure 11.2(b), there is a longer progression which shows an intensity maximum, in this case at $v' = 4$; this is shown in Figure 11.3(b).

The example in Figure 11.3(b) shows how electronic absorption spectroscopy of diatomic molecules can allow access to many vibrational levels of an excited electronic state. This is dependent on $r_e$ in the excited state being appreciably different from that in the ground state. For example, $r_e$ is 0.2666 and 0.3025 nm (2.666 and 3.025 Å) in the ground and first (triplet) excited states, respectively, of $I_2$ (the absorption spectrum which gives iodine vapour its characteristic violet colour). In this

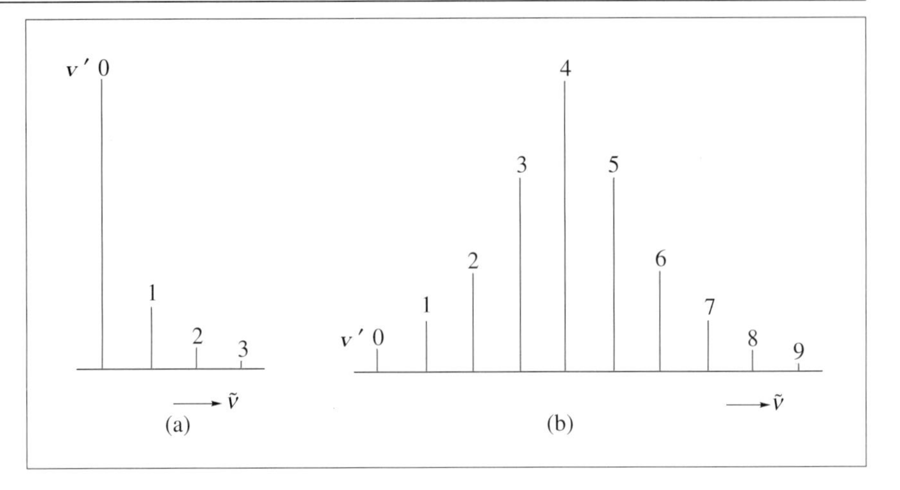


**Figure 11.3** Typical intensity distribution along a vibrational progression when (a) $r_e' \approx r_e''$ and (b) $r_e' > r_e''$

case, the increase of bond length in the excited state is so large that the maximum intensity in the progression from $v'' = 0$ is at about $v' = 30$, and the 0–0 transition is so weak that it has not been observed. From a large number of observed vibrational levels it is possible to construct the vibrational potential function and to determine the dissociation energy, $D_0$ or $D_e$, in an excited electronic state.

In electronic emission spectra, whether **fluorescence**, with $\Delta S = 0$, where $S$ is the electron spin quantum number, or **phosphorescence**, with $\Delta S = \pm 1$, corresponding information relating to the ground electronic state may be obtained. Again, the length of an observed progression depends on the difference in $r_e$ between the two electronic states. If it is allowed by the selection rules, an emission spectrum may be observed between a pair of excited electronic states.

If the difference in bond length between the ground and excited electronic states is sufficiently large, the $v'' = 0$ progression in an absorption spectrum may extend up to, and beyond, the excited state dissociation limit.

Accurate measurements of the intensity distribution along a vibrational progression may be used to calculate a value of $\Delta r$, the change of bond length between two electronic states. However, we shall see in Section 11.2.3 that this is more readily obtained from the rotational fine structure accompanying the electronic transition.

In Section 9.2.4 it was explained that the $v'' = 1$ level, and even higher vibrational levels in the ground electronic state, may be appreciably populated if the level is sufficiently low-lying (or if the temperature is sufficiently high). In this case, **sequence bands**, groups of bands with constant $\Delta v$, may be observed in the electronic absorption spectrum. The sequences with $\Delta v = 0$ and 1 are illustrated in Figure 11.4.

Such sequences consist of closely spaced groups of bands. The spacings within a sequence result from the difference in vibration wave-

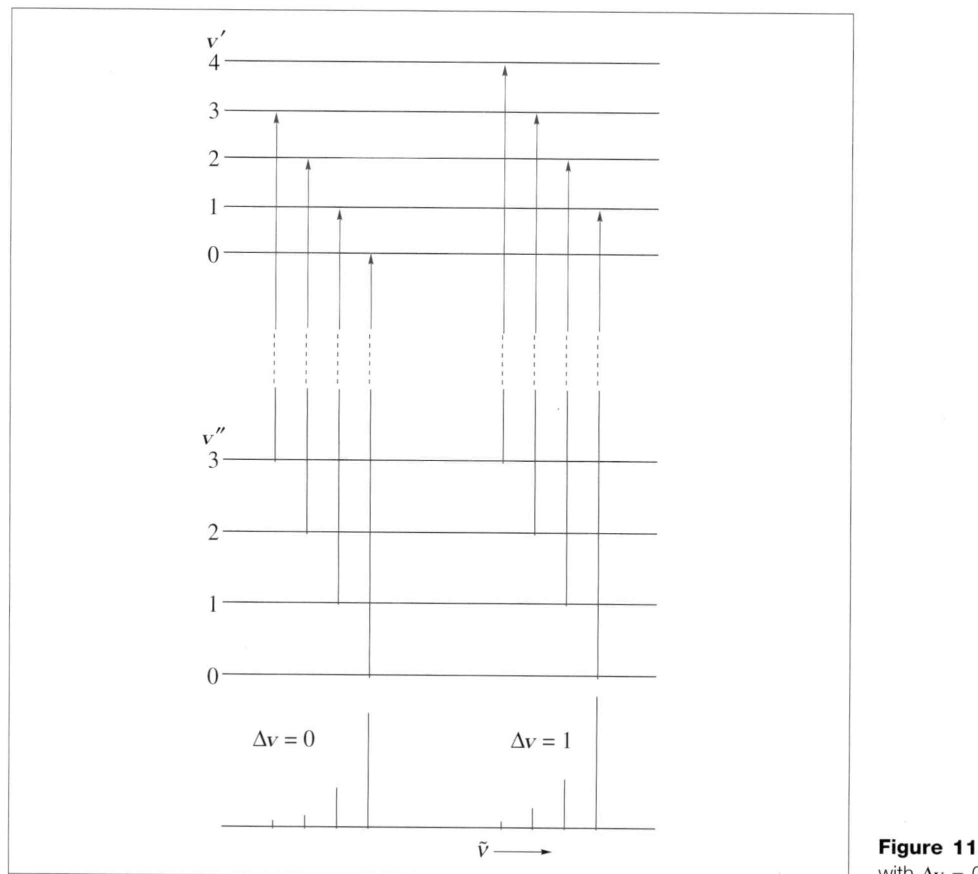


**Figure 11.4** Sequence bands with $\Delta v = 0$ and 1

numbers in the two electronic states. In the case illustrated, the vibration wavenumber is higher in the ground state than in the excited state, leading to sequence bands being on the low wavenumber side of the main bands. Sequence band spacings are small compared to the separations of bands in a progression.

### 11.2.3 Rotational Fine Structure

Typical potential energy curves for the ground and an excited electronic state of a diatomic molecule were shown in Figure 6.9. The figure shows the vibrational energy levels associated with each state and indicates also that there is a stack of rotational energy levels associated with each of these. Consequently, fine structure, consisting of rotational transitions, accompanies all electronic and vibronic transitions.

The rotational fine structure has much in common with that found in vibrational spectra, discussed in Section 10.2. The main difference concerns the change of the rotational constant, $B$, from the lower to the upper state of a transition. The constants $B_0''$ and $B_1''$, for the $v'' = 0$ and

1 levels of the ground electronic state, are very similar, *i.e.* $\alpha$ in equation (10.4) is very small. On the other hand, there is typically a relatively large change of $B$ from one electronic state to another. In forming an excited electronic state it is often the case that an electron is promoted from a bonding to an antibonding (or, at least, less strongly bonding) orbital, as discussed in Sections 5.1.2 and 5.2.2. This causes an increase in bond length and a decrease in the value of $B$, which is typically a much larger change than between $B_0''$ and $B_1''$.

During an electronic transition there is a change of dipole moment, similar in principle to the change of dipole moment when an infrared vibrational transition takes place, as discussed in Section 10.4. Associated with this is a transition moment, $\boldsymbol{R}_e$, and the transition intensity is proportional to $\boldsymbol{R}_e^2$. As for a vibration of a linear polyatomic molecule (see Section 10.4), this transition moment may be polarized along the internuclear axis, resulting in a parallel band, or perpendicular to it, to give a perpendicular band. We shall consider the rotational fine structure associated with only two types of electronic transition, $^1\Sigma^+$–$^1\Sigma^+$ and $^1\Pi$–$^1\Sigma^+$, in heteronuclear diatomic molecules. The rotational fine structure is the same for $^1\Sigma_u^+$–$^1\Sigma_g^+$ and $^1\Pi_u$–$^1\Sigma_g^+$ transitions in homonuclear diatomic molecules. It is also the same for all members of vibrational progressions and sequences built on these electronic transitions.

**Box 11.2 Electronic and Vibronic Transition Moments**

The transition moment, $\boldsymbol{R}_e$, for a transition between lower and upper states with electronic wave functions $\psi_e''$ and $\psi_e'$, respectively, is given by:

$$\boldsymbol{R}_e = \int \psi_e' \boldsymbol{\mu} \psi_e'' \mathrm{d}\tau_e \qquad (11.7)$$

where $\boldsymbol{\mu}$ is the dipole moment and the integration is over electronic coordinates $\tau_e$.

For a vibronic transition, the transition moment, $\boldsymbol{R}_{ev}$, between lower and upper states with vibronic wave functions $\psi_{ev}''$ and $\psi_{ev}'$, respectively, is given by:

$$\boldsymbol{R}_e = \int \psi_{ev}' \boldsymbol{\mu} \psi_e'' \mathrm{d}\tau_e \qquad (11.8)$$

A vibronic wave function, $\psi_{ev}$, can be factorized into $\psi_e \psi_v$, so that:

$$\boldsymbol{R}_{ev} = \int \psi_e' \boldsymbol{\mu} \psi_e'' \mathrm{d}\tau_e \int \psi_v' \psi_v'' \mathrm{d}\tau_v = \boldsymbol{R}_e \int \psi_v' \psi_v'' \mathrm{d}\tau_v \qquad (11.9)$$

This is the product of the electronic transition moment, which is assumed to be constant for all vibronic transitions, and the vibrational **overlap integral**. The overlap integral is a measure of the extent to which the vibrational wave functions in the two electronic states overlap. For example, the transitions indicated in Figure 11.2(a) and 11.2(b) correspond to maximum overlap of the vibrational wave functions for $v'' = 0$ and $v' = 0$ in Figure 11.2(a), and for $v'' = 0$ and $v' = 4$ in Figure 11.2(b).

The intensities of electronic and vibronic transitions are proportional to $\boldsymbol{R}_e^2$ and $\boldsymbol{R}_{ev}^2$, respectively.

### $^1\Sigma^+$–$^1\Sigma^+$ (or $^1\Sigma_u^+$–$^1\Sigma_g^+$) Electronic or Vibronic Transition

For this type of transition the transition moment is polarized along the internuclear axis. The rotational selection rule is:

$$\Delta J = \pm 1 \qquad (11.10)$$

This is the same rule as for rotational transitions accompanying a vibrational infrared transition, as given in equation (10.1), and results in an R-branch, with $\Delta J = +1$, and a P-branch, with $\Delta J = -1$. The energy level diagram and the allowed transitions shown in Figure 10.1 apply also to a $^1\Sigma^+$–$^1\Sigma^+$ electronic transition, as indicated by the symbols on the right-hand side of the figure.

Figure 11.5 shows the rotational fine structure of the $^1\Sigma^+$–$^1\Sigma^+$ pure electronic transition, from $v'' = 0$ to $v' = 0$ (the 0–0 band), of the short-lived molecule RhN (rhodium mononitride). Although the rotational selection rules are the same as for the $v = 1$–0 transition of HCl, shown in Figure 10.2, the spectrum of RhN appears significantly different. However, the R-branch is still on the high wavenumber (low wavelength) side and the P-branch on the low wavenumber (high wavelength) side of the zero gap. The zero gap is the comparatively wide spacing between the first line of the P-branch, P(1), and the first line of the R-branch, R(0). Although the spectrum of HCl is slightly unsymmetrical, showing some convergence in the R-branch and divergence in the P-branch, that of RhN is very unsymmetrical: there is much greater convergence in the R-branch and divergence in the P-branch. The convergence in the R-branch is so extreme that it converges to a **band head**, where several lines are almost coincident, and then reverses, continuing to low wavenumber. In fact, two weak lines of the reversed R-branch are clearly visible in the zero gap.

Irrespective of the degree of asymmetry of the band, the method sum-

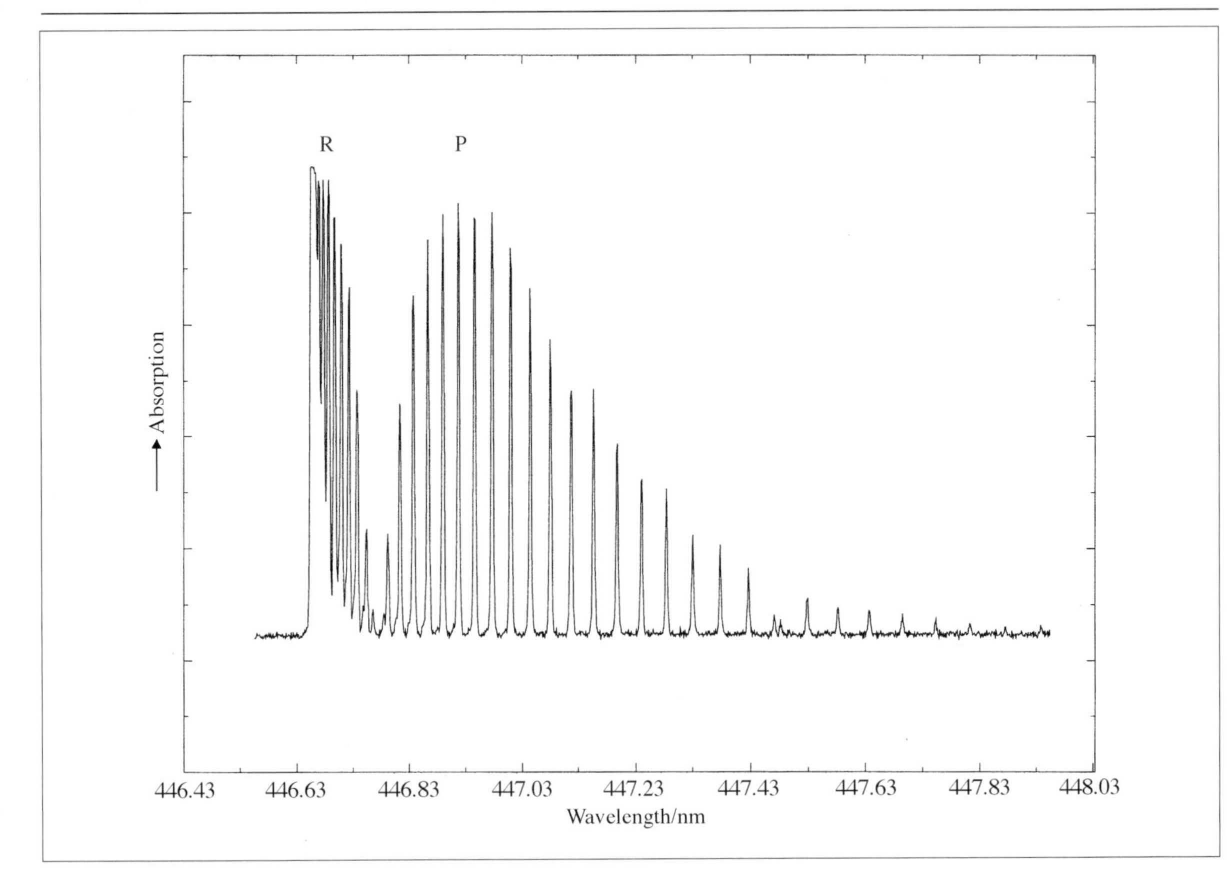


**Figure 11.5** Rotational fine structure of a $^1\Sigma^+$–$^1\Sigma^+$ electronic transition of RhN

marized in equations (10.7) and (10.8) may be used to obtain the *B*-value for the upper and lower states, respectively. For the 0–0 band of an electronic spectrum, $\omega_0$ is the wavenumber of the pure electronic transition, and these equations become:

$$\tilde{\nu}[\mathrm{R}(J)] - \tilde{\nu}[\mathrm{P}(J)] = 4B_0'(J + \tfrac{1}{2}) \qquad (11.11)$$

and

$$\tilde{\nu}[\mathrm{R}(J-1)] - \tilde{\nu}[\mathrm{P}(J+1)] = 4B_0''(J + \tfrac{1}{2}) \qquad (11.12)$$

where $B_0'$ and $B_0''$ are the *B*-values for the zero-point levels of the upper and lower electronic states, respectively. Plotting values of $\tilde{\nu}[\mathrm{R}(J)] - \tilde{\nu}[\mathrm{P}(J)]$ against $(J + \tfrac{1}{2})$ gives a straight line of slope $4B_0'$, and plotting $\tilde{\nu}[\mathrm{R}(J-1)] - \tilde{\nu}[\mathrm{P}(J+1)]$ against $(J + \tfrac{1}{2})$ gives a straight line of slope $4B_0''$.

We saw in Section 11.2.2 that vibrational progressions may be observed if the Franck–Condon principle confers sufficient intensity. In an absorption spectrum, this allows access to more vibrational levels in the excited electronic state. All the bands in a progression associated with a $^1\Sigma$–$^1\Sigma$ type of electronic transition have the same kind of rotational fine structure as the 0–0 band. The only difference is caused by $B'$ varying with

**Worked Problem 11.2**

**Q** Using the wavelength scale on Figure 11.5, measure, as accurately as you can, the wavelengths of the R(5), P(5) and P(7) lines of the spectrum of RhN (at best, these wavelengths should be accurate to about ±0.001 nm, but using a photocopier to double the size of the figure will improve the accuracy). Convert these to wavenumbers, and use equations (10.7) and (10.8) to obtain approximate values for $B_0'$ and $B_0''$, respectively.

**A** For R(5), $\lambda = 446.6866$ nm (retaining the fourth decimal place)
$\therefore \tilde{\nu} = 22\ 387.06\ \text{cm}^{-1}$
For P(5), $\lambda = 446.8900$ nm
$\therefore \tilde{\nu} = 22\ 376.87\ \text{cm}^{-1}$
For P(7), $\lambda = 446.9483$ nm
$\therefore \tilde{\nu} = 22\ 373.95\ \text{cm}^{-1}$
$\therefore \tilde{\nu}[\text{R}(5)] - [\text{P}(5)] = 10.19\ \text{cm}^{-1} = 4B_0' \times 5.5$
$\therefore B_0' = 0.463\ \text{cm}^{-1}$
and $[\text{R}(5)] - [\text{P}(7)] = 13.11\ \text{cm}^{-1} = 4B_0'' \times 6.5$
$\therefore B_0'' = 0.504\ \text{cm}^{-1}$

These values are, perhaps surprisingly, good compared to the published (W. J. Balfour and S. Fougère, *J. Mol. Spectrosc.*, 2000, **199**, 18) values of $B_0'' = 0.5073\ \text{cm}^{-1}$ and $B_0' = 0.4634\ \text{cm}^{-1}$, obtained from the wavenumbers of all the P- and R-branch lines.

$v'$ in exactly the same way (see equation 10.4) as it does in the ground electronic state:

$$B_v' = B_e' - \alpha'(v' + \tfrac{1}{2}) \qquad (11.13)$$

Determination of as many values as possible of $B_v'$ gives values of the vibration–rotation interaction constant, $\alpha'$, and $B_e'$, from which the equilibrium bond length, $r_e'$, follows.

Similarly, observation of the rotational fine structure of members of a progression in an electronic emission spectrum to the ground electronic state gives values of $B_v''$. Then, since:

$$B_v'' = B_e'' - \alpha''(v'' + \tfrac{1}{2}) \qquad (11.14)$$

$\alpha''$, $B_e''$ and therefore $r_e''$ can be determined.

We have seen in equation (10.4) that $\alpha''$ and $B_e''$ can be determined also from the vibration–rotation spectrum, usually from the observation of only $B_0''$ and $B_1''$. Determination of $\alpha''$ and $B_e''$ from values of $B_v''$ obtained from the members of a vibrational progression, preferably a lengthy one, in an electronic emission spectrum results in more accurate values for these parameters.

**Worked Problem 11.3**

**Q** For the ground electronic state of $^{19}F_2$, $B_0'' = 0.883\ 26\ cm^{-1}$ and $B_1'' = 0.869\ 41\ cm^{-1}$. Calculate the value of $\alpha''$, and the bond length, $r_e''$, given that $m(^{19}F) = 18.998$ u.

**A** From equation (11.14):

$0.883\ 26\ cm^{-1} = B_e'' - 0.5\alpha''$ (1)

$0.869\ 41\ cm^{-1} = B_e'' - 1.5\alpha''$ (2)

Then, (1) – (2) gives $\alpha = 0.013\ 85\ cm^{-1}$, from which it follows that $B_e'' = 0.890\ 19\ cm^{-1}$.

From equation (7.3), $\mu = \frac{1}{2}m(^{19}F) = 0.5 \times 18.998 \times 1.6605 \times 10^{-27}$ kg $= 1.5773 \times 10^{-26}$ kg

From equation (9.3):

$$r_e''^2 = h/8\pi^2 c\mu B_e''$$
$$= 6.6261 \times 10^{-34}\ \text{J s}/8\pi^2 \times 2.9979 \times 10^{10}\ \text{cm s}^{-1} \times 1.5773 \times 10^{-26}\ \text{kg} \times 0.890\ 19\ \text{cm}^{-1}$$
$$= 1.9937 \times 10^{-20}\ \text{m}^2$$
$$\therefore r_e'' = 1.4120 \times 10^{-10}\ \text{m} = 1.4120\ \text{Å or } 0.141\ 20\ \text{nm}$$

### $^1\Pi$–$^1\Sigma^+$ (or $^1\Pi_u$–$^1\Sigma_g^+$) Electronic or Vibronic Transition

For this type of transition, the transition moment is polarized perpendicular to the internuclear axis. The rotational selection rule is:

$$\Delta J = 0, \pm 1 \qquad (11.15)$$

and is the same as for the rotational transitions accompanying a vibrational infrared transition of a linear polyatomic molecule, as in equation (10.12). The energy level diagram for a $^1\Pi$–$^1\Sigma^+$ (or $^1\Pi_u$–$^1\Sigma_g^+$) transition is like that in Figure 10.6, but using the symbols on the right-hand side of the figure. Just as for a vibrational transition, there are P-, Q- and R-branches. The rotational levels of the upper, $^1\Pi$, state start with $J = 1$, and they are all doubled.

For a $\Pi$ electronic state, the quantum number $\Lambda$, associated with the component of the orbital angular momentum along the internuclear axis, is 1 (see equation 5.10). Such a state is doubly degenerate, which can be thought as being due to the same energy being associated with clockwise or anticlockwise motion of the electrons around the internuclear axis. However, interaction between the orbital motion of the electrons and the rotation of the molecule causes a splitting, $\Delta F(J)$, of the rotational levels given by:

$$\Delta F(J) = qJ(J + 1) \tag{11.16}$$

where $q$ is constant for a particular electronic state. The effect is known as **$\Lambda$-type doubling**. The doubling increases with $J$, as shown by equation (11.16) and Figure 10.6, but whereas the doubling is typically very small for a vibrational state, it is more likely to be appreciable for an electronic state.

Figure 11.6 shows the rotational fine structure of the 1–0 vibronic transition associated with a $^1\Pi$–$^1\Sigma^+$ electronic transition of $^{187}$ReN (rhenium mononitride). The band shows typical P-, Q- and R-branch structure and is slightly complicated by the presence of a similar band due to the isotopic species $^{185}$ReN displaced by 0.33 cm$^{-1}$ to high wavenumber. There is convergence in the R-branch and divergence in the P-branch, owing to a decrease of $B$ (an increase in the bond length) from the ground to the excited state, although this is less pronounced than in the spectrum

$^{187}$Re and $^{185}$Re have natural abundances of 62.6% and 37.4%, respectively.

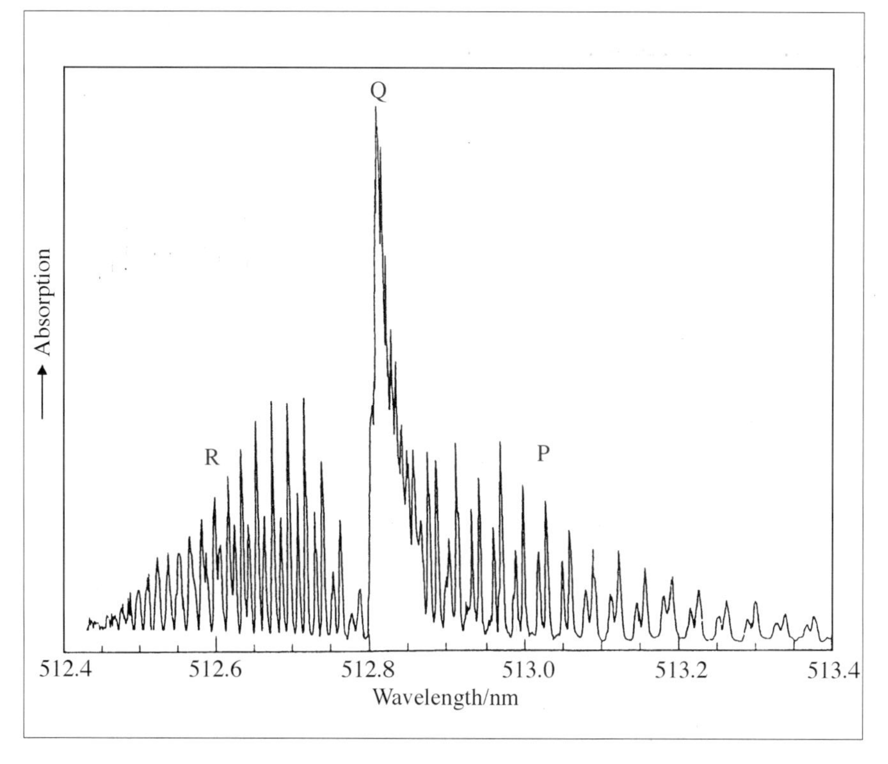


**Figure 11.6** Rotational fine structure of a $^1\Pi$–$^1\Sigma^+$ vibronic transition of $^{187}$ReN (and $^{185}$ReN)

of RhN in Figure 11.5. The slight divergence of the Q-branch to low wavenumber is also due to the decrease of the value of $B$.

One result of the splitting of all the rotational levels in the $^1\Pi$ state is that there are two effective $B$-values, one for the lower set of levels, $B_{0(l)}'$, and one for the upper, $B_{0(u)}'$. Figure 10.6, indicating that the P- and R-branch transitions go to the lower set, shows that equation (11.10) can be used, in the same way as for a $^1\Sigma$–$^1\Sigma$ transition, by plotting $\tilde{\nu}[\mathrm{R}(J)] - \tilde{\nu}[\mathrm{P}(J)]$ against $(J + \frac{1}{2})$, to obtain $B_{0(l)}'$ from the wavenumbers of these transitions. $B_{0(u)}'$ can be obtained from the Q-branch transitions, and $q$, in equation (11.16), follows from $B_{0(l)}'$ and $B_{0(u)}'$. From the spectrum of $^{187}$ReN shown in Figure 11.6 it was found that $B_1' = 0.466\ 63\ \mathrm{cm}^{-1}$ and $q_1' = -0.000\ 13\ \mathrm{cm}^{-1}$.

These values of $B_1'$ and $q_1'$ are for the $v = 1$ vibrational level of the $^1\Pi$ electronic state. Irregularities in the rotational structure of the 0–0 band in this electronic band system of ReN precluded a determination of the values of $B_0'$ and $q_0'$.

The value of $B_0'' = 0.481\ 122\ 3\ \mathrm{cm}^{-1}$, for the $^1\Sigma^+$ ground electronic state, had been determined from the 0–0 band of another electronic transition.

## 11.3 Electronic Spectra of Polyatomic Molecules

### 11.3.1 Vibrational Coarse Structure

It was shown in Section 6.3 that linear and non-linear polyatomic molecules have $3N - 5$ and $3N - 6$ normal vibrations, respectively, where $N$ is the number of atoms in the molecule. In electronic spectra of these molecules the dominant active vibrations are usually those which retain the symmetry of the molecule, so-called **totally symmetric vibrations**. The selection rule which applies to each of these vibrations is the same as for a diatomic molecule, namely:

$$\Delta v = 0, \pm 1, \pm 2, \pm 3, \ldots \qquad (11.17)$$

as in equation (11.6). The selection rule is unrestrictive but, as in the electronic spectra of diatomic molecules, the Franck–Condon principle (explained in Section 11.2.2) applies. In polyatomic molecules the Franck–Condon principle applies to each vibration, separately. If there is a change of molecular geometry from the ground (or lower) to an excited (or upper) electronic state, vibrational progressions are observed in any vibrations which tend to take the molecule into its new geometry.

In Section 5.3.1 a useful shorthand labelling for electronic states of polyatomic molecules was explained. The symbols and S and T refer to singlet and triplet states, respectively. $S_0$ and $S_1$ are the ground and first excited singlet states, respectively.

There is a clear example of the Franck–Condon principle applying to an electronic spectrum of a polyatomic molecule in the $S_1$–$S_0$ absorption spectrum of benzene. This occurs in the near-ultraviolet region and is shown, in the gas phase at low resolution, in Figure 11.7. The transition involves the promotion of an electron from the $\pi_2,\pi_3$ pair to the $\pi_4,\pi_5$ pair of orbitals shown in Figure 5.13. In comparison with the former pair, the latter pair show an additional nodal plane perpendicular to the

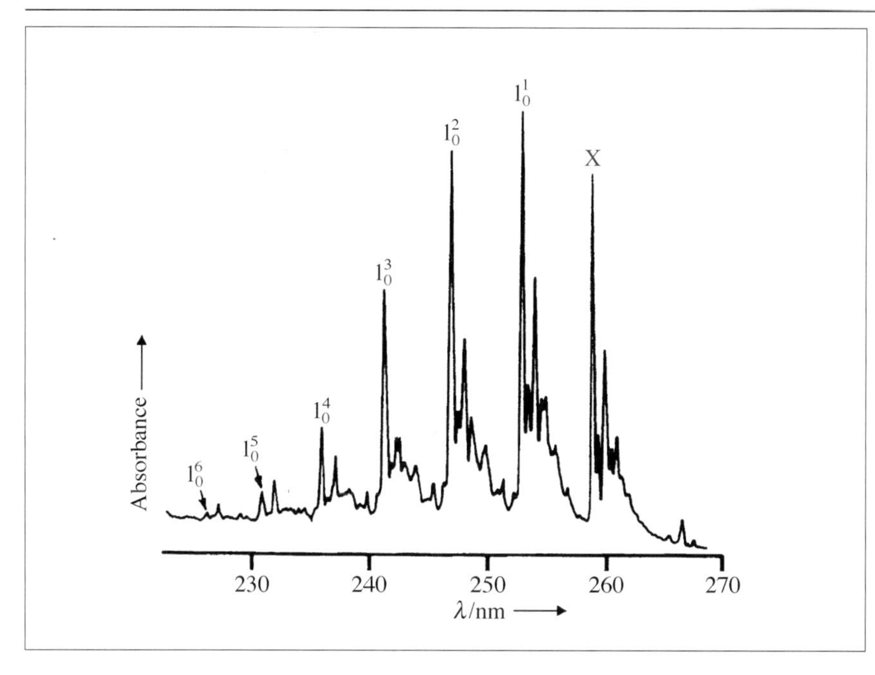


**Figure 11.7** Low-resolution, near-ultraviolet absorption spectrum of benzene in the gas phase (after J. M Hollas, *Modern Spectroscopy*, 3rd edn., Wiley, New York, 1996, p. 245)

molecule, resulting in a uniform expansion of the carbon ring from $S_0$ to $S_1$. In fact, all the C–C bond lengths increase from 0.1397 nm (1.397 Å) in $S_0$ to 0.1434 nm (1.434 Å) in $S_1$. The vibration which tends to take the molecule into this $S_1$ geometry is the so-called ring-breathing vibration, labelled $\nu_1$, a vibration in which the carbon ring breathes symmetrically in and out from its conformation in $S_0$. The progression, built on the band labelled X, shows six members, labelled $1_0^n$, where $n$ = 1–6.

It might be expected that the band X would be the pure electronic transition, the 0–0 band, as in the spectrum of a diatomic molecule. In a highly symmetrical molecule, such as benzene, it is possible that an electronic band system may be observed in spite of the fact that it is forbidden by the electronic selection rules. In such a case the electronic transition is induced by a vibration of the appropriate symmety. In the case of the $S_1$–$S_0$ transition in benzene, it is induced by the vibration $\nu_6$ so that band X is a vibronic transition from the zero-point level of $S_0$ to one quantum of $\nu_6$ in $S_1$.

On the high-wavelength side of each member of the progression in Figure 11.7 are a series of weaker, sequence bands. Just like sequence bands in electronic spectra of diatomic molecules, discussed in Section 11.2.2 and illustrated in Figure 11.4, they are bands with $\Delta v = 0$ originating in vibrational levels in the ground electronic state which are significantly populated at ambient temperature. The Boltzmann distribution of molecules among these levels is given by equation (9.4). The lowest wavenumber vibrations have the highest populations and therefore are responsible for the most intense sequences.

The electronic spectra of large molecules with lower symmetry than benzene show more complex vibrational structure. This is partly due to the fact that more vibrations retain the symmetry of the molecule and are allowed to form progressions. It is also due to the fact that a geometry change in the excited electronic state may involve more than one normal vibration. For example, in fluorobenzene there is a general expansion of the benzene ring, similar to that in benzene itself, and a small contraction of the C–F bond in the $S_1$ state. There are now several

In general, the bond lengths in molecules in excited electronic states tend to be larger than in the ground state. The reason for this is that the electronic transition often involves the promotion of an electron from a bonding, or non-bonding, orbital to an orbital with more anti-bonding characteristics. Consequently, force constants and vibration wavenumbers tend to decrease (see equation 6.8) in the excited electronic state. The application of Figure 11.4 to any vibration responsible for a sequence shows that, in this case, the sequence bands are situated to low wavenumber (high wavelength) of the band with which they are associated. In the more unusual cases, where the orbital into which the electronic is promoted is more strongly bonding, sequences may be observed on the high wavenumber (low wavelength) side of the main band.

vibrations which, although retaining the symmetry of the molecule, involve some ring breathing or C–F stretching character. According to the Franck–Condon principle, each of these may form a progression.

In a large molecule the various possible progressions and numerous sequences result in a very complex, and crowded, gas-phase absorption spectrum.

### 11.3.2 Rotational Fine Structure

Rotational fine structure of electronic and vibronic transitions in polyatomic molecules is a vast subject, which we shall only touch on here.

Linear polyatomic molecules show rotational structure in their electronic spectra which is closely related to that of diatomic molecules. The structure associated with $^1\Sigma$–$^1\Sigma$ and $^1\Pi$–$^1\Sigma$ types of transitions is of the same type as for a diatomic molecule; the former show a P- and R-branch and the latter a P-, Q- and R-branch. The main difference is a tendency for greater congestion of rotational transitions for longer and heavier molecules. There is a problem in respect of determination of molecular structure in that there is only one rotational constant, $B$, for any electronic state, which gives only one structural parameter. If several bond lengths are required, isotopic substitution, where possible, is necessary (see Section 9.4).

Electronic and vibronic transitions in symmetric top molecules give rise to parallel or perpendicular bands, in which the electronic or vibronic transition moment (see Box 11.2) may be parallel or perpendicular to the unique axis of the molecule (see Section 10.4 for a discussion of parallel and perpendicular bands in the vibrational spectra of symmetric tops): the $a$-axis in a prolate or the $c$-axis in an oblate symmetric top (see Section 7.3.2). The rotational selection rules are the same as for a parallel or perpendicular band in the infrared vibrational spectrum of a symmetric top. These are given in equations (10.13) and (10.14) for parallel and perpendicular bands, respectively.

Although the selection rules for parallel and perpendicular bands of symmetric tops are the same as for infrared vibrational spectra, the rotational fine structure of electronic spectra usually has a significantly different general appearance. The reason for this is that the changes of rotational constants, $A$ and $B$ for a prolate and $C$ and $B$ for an oblate symmetric top, are typically much larger than for a vibrational transition. In addition, these changes vary from one electronic transition to another, even in the same molecule. We can no longer think of a typical parallel or perpendicular band in the electronic spectrum of a symmetric top. As for linear molecules, there is an increasing congestion of rotational transitions for larger and heavier symmetric tops.

The complex energy levels and selection rules that apply to infrared vibrational spectra of asymmetric top molecules (see Section 10.6) apply also to their electronic spectra. In a molecule of sufficiently high symmetry, the electronic (or vibronic) transition moment may be polarized along the *a*, *b* or *c* inertial axis, resulting in a type A, B or C band, respectively. In an asymmetric top molecule of lower symmetry, bands may be mixtures of these types, so-called **hybrid bands**

Whereas, in infrared vibrational spectra, changes of rotational constants, *A*, *B* and *C*, from the lower to the upper state are typically very small, resulting in bands having a symmetrical appearance, changes in electronic spectra are usually relatively large, resulting in bands having an asymmetrical shape. We can no longer think of typical type A, B or C bands, as we could for infrared vibrational spectra (see, for example, the vibrational spectrum of fluorobenzene in Figure 10.10).

For relatively large molecules the rotational fine structure in the electronic spectra of asymmetric tops may be so congested that all we are able to observe is a contour of the underlying, unresolved rotational structure. This was the case also for the vibrational spectrum of fluorobenzene in Figure 10.10. Figure 11.8 shows type A, B and C bands for aniline for the $S_1$–$S_0$ electronic transition. The spectrum in Figure 11.8(a) shows the rotational contour of the 0–0 band, observed in the absorption spectrum and containing tens of thousands of unresolved rotational transitions. That in Figure 11.8(b) shows a contour which was computed with values of rotational constants, *A*, *B* and *C*, which produce the best agreement with the observed contour. The contours in Figure 11.8(c) and 11.8(d) have been computed with the same rotational constants, but with type A and C selection rules, respectively. Although no type A or C bands have been observed in this electronic spectrum, the computed contours serve to show how different are the contours of the three band types. It is also informative to note how different these contours are from those in the infrared vibrational spectrum of fluorobenzene in Figure 10.10. Although the ground state constants $A''$, $B''$ and $C''$ are very similar for both molecules, it is the much larger changes in the electronic excited state of aniline than occur in the vibrationally excited states of fluorobenzene which produce the considerable differences between the contours in Figures 10.10 and 11.8.

### 11.3.3 Electronic Spectra of Molecules in Solution

An example of an electronic spectrum of a molecule in solution is that of *trans*-dimethyldiimide [(*E*)-1,2-dimethyldiazene] in hexane, shown in Figure 8.4. Typical of many such spectra, each electronic band system, centred here at about 28 200 ($S_2$–$S_0$) and 42 700 cm$^{-1}$ ($S_1$–$S_0$), shows no coarse vibrational structure. The vibrational energy levels are broadened to such an extent, by energy exchange due to collisions with the solvent

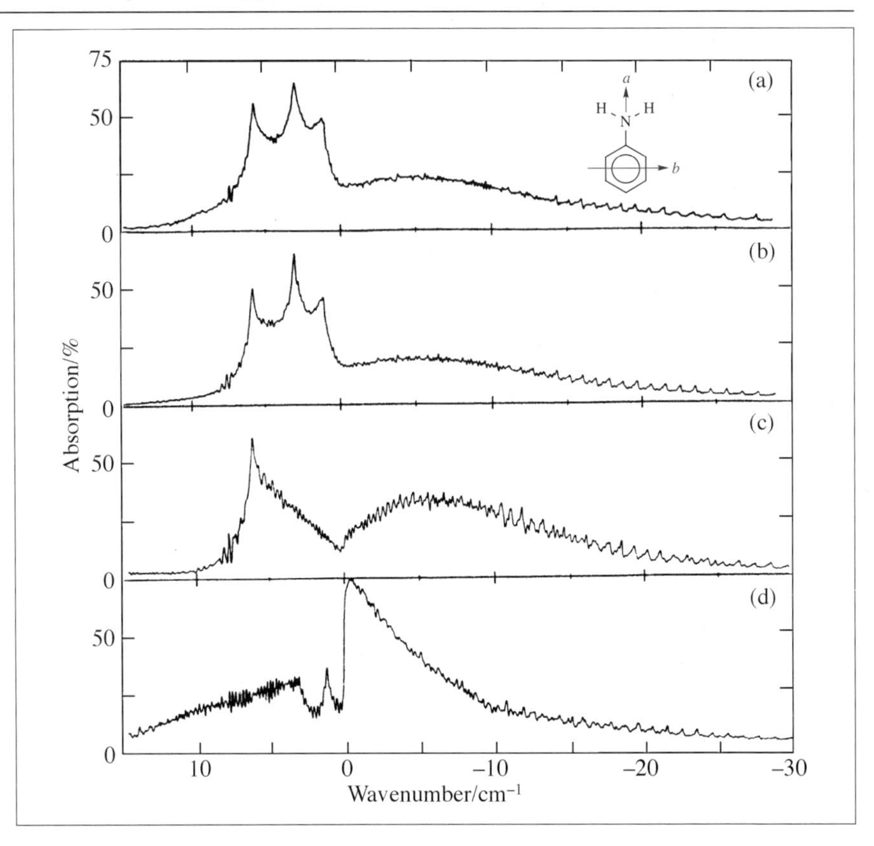


**Figure 11.8** (a) Observed type B and computed (b) type B, (c) type A and (d) type C rotational contours for aniline, using rotational constants for the 0–0 band (after J. Christoffersen, J. M. Hollas and G. H. Kirby, *Mol. Phys.*, 1969, **16**, 448)

molecules, that they merge together. Nevertheless, as was discussed in Section 8.1, such spectra are very useful in chemical analysis. If the molar absorption coefficient, $\varepsilon$, is known at a particular wavelength, measurement of the absorbance, $A$, at that wavelength gives the concentration, $c$ (see equation 8.2).

This is probably the most important use of electronic spectra in analytical chemistry. However, electronic spectra in solution are used also to indicate the presence of groups within a molecule. For example, a carbonyl (C=O) group or a benzene ring confers absorption at a characteristic wavelength and intensity. Such a group is called a **chromophore**. In Figure 8.4, for example, the weak absorption at 28 200 $cm^{-1}$ (354.6 nm) is characteristic of the N=N chromophore.

Literally, chromophore means "conferring a particular colour" on a molecule. Using "colour" in the most general sense, a chromophore confers absorption of a particular wavelength on the molecule; this wavelength may or may not be in the visible region of the spectrum.

The lowest energy, $S_1$–$S_0$, electronic transition of ethene ($CH_2$=$CH_2$) shows a broad absorption centred at about 162 nm. The transition involves an electron promotion from the $\pi$ to the $\pi^*$ orbital, shown in Figure 5.10, to give a $\pi\pi^*$ $S_1$ state, as discussed in Section 5.3.1. This intense absorption is characteristic of molecules containing an electronically isolated C=C group. If, however, the C=C group is conjugated to another CC multiple bond, as in buta-1,3-diene discussed in Section 5.3.3, there is a considerable shift to high wavelength of the $S_1$–$S_0$ transition. The $S_1$ state in buta-1,3-diene is also a $\pi\pi^*$ state in which an electron is promoted from the $\pi_2$ to the $\pi_3$ orbital, shown in Figure 5.12. The absorption spectrum of penta-1,3-diene in solution in heptane (Figure 11.9) shows that the effect of conjugation of the two double bonds is to shift the $S_1$–$S_0$ system from 162 nm in ethene to 224 nm. In fact, it is a quite general observation that the greater is the delocalization of the $\pi$ orbitals in a molecule, the more is the $S_1$–$S_0$ system shifted to low energy (high wavelength).

The presence of the methyl group in penta-1,3-diene results in only a very small shift of the $S_1$–$S_0$ system from that of buta-1,3-diene.

A characteristic of a transition to a $\pi\pi^*$ state is a large value of $\varepsilon_{max}$, resulting in a high intensity. In the 224 nm system of penta-1,3-diene, $\varepsilon_{max} = 2.6 \times 10^4$ mol$^{-1}$ dm$^3$ cm$^{-1}$.

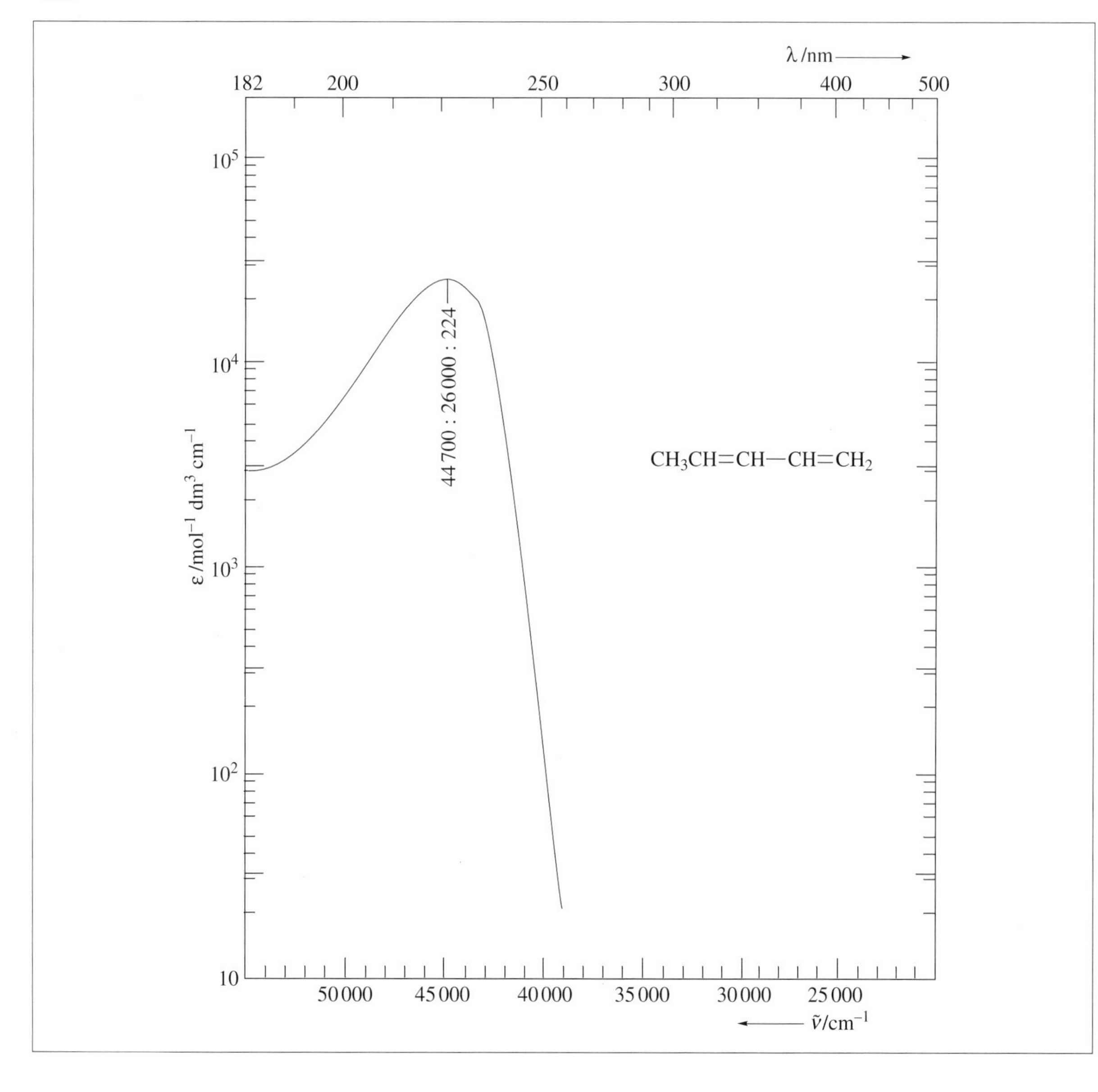


**Figure 11.9** Near-ultraviolet absorption spectrum of penta-1,3-diene in solution in heptane (after H.-H. Perkampus, in *UV Atlas of Organic Compounds*, Butterworths, London, 1966, p. A3/1)

Figure 11.10 shows the electronic absorption spectrum of benzene in solution in hexane. The shift of the $S_1$–$S_0$ system to about 260 nm, compared to 224 nm for penta-1,3-diene, is caused by the greater degree of π-electron delocalization in benzene. The much lower intensity ($\varepsilon_{max}$ = 250 mol$^{-1}$ dm$^3$ cm$^{-1}$) is due to the fact that, although $S_1$ is a ππ* state, the electronic transition is forbidden and appears only by virtue of being vibrationally induced, as discussed in Section 11.3.1. The $S_1$–$S_0$ system, and the $S_2$–$S_0$ system at about 204 nm, involve promotion of an electron from the $\pi_2\pi_3$ to the $\pi_4\pi_5$ pairs of orbitals in Figure 5.13.

The spectrum in Figure 11.10 is rather unusual in that, although it is

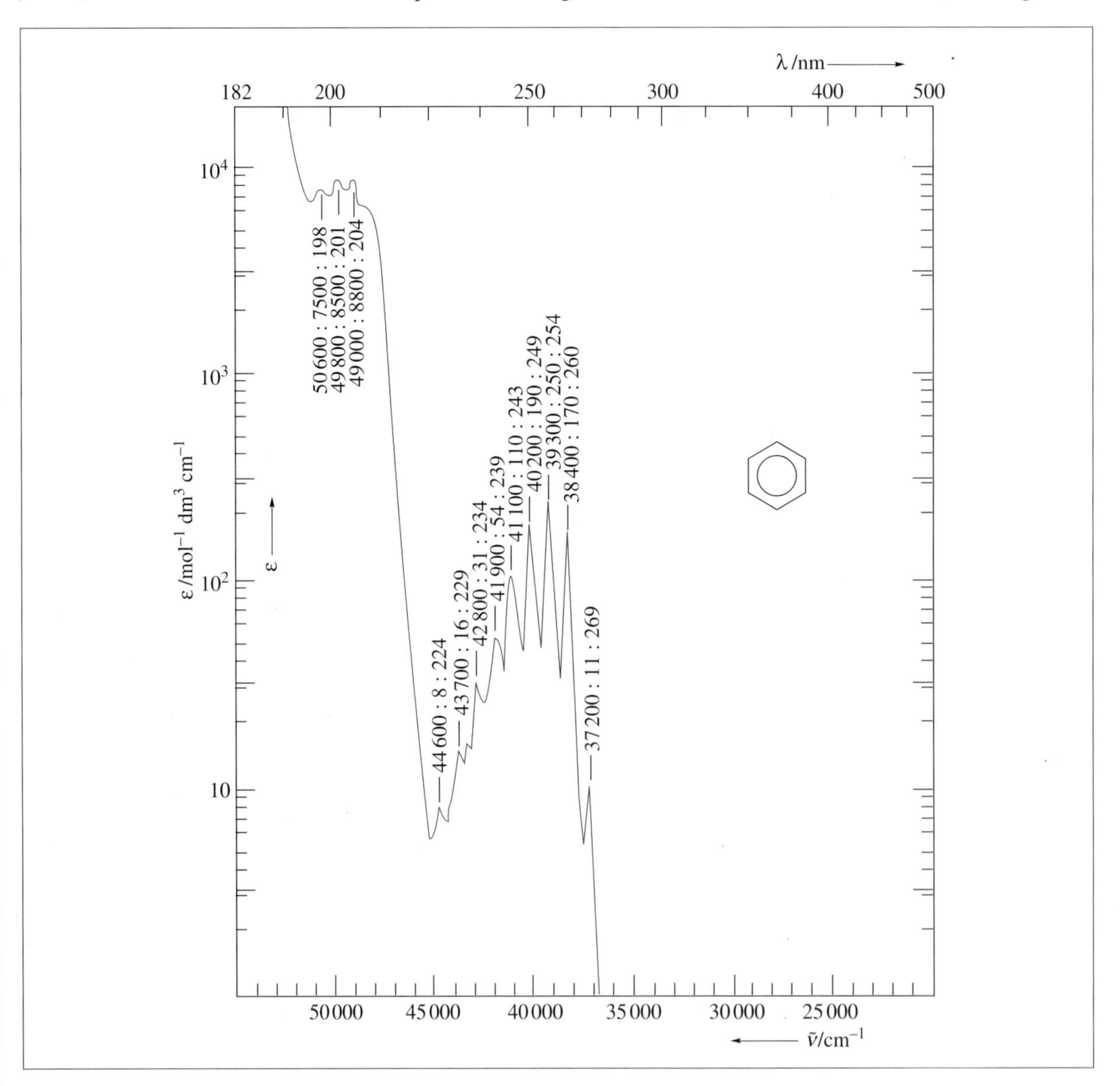


**Figure 11.10** Near-ultraviolet absorption spectrum of benzene in solution in hexane (after H.-H. Perkampus and G. Kassebeer, in *UV Atlas of Organic Compounds*, Butterworths, London, 1966, p. D1/1)

in solution, it shows considerable vibrational structure (compare the gas-phase spectrum in Figure 11.7). In general, there is a greater tendency for vibrational structure to be observed in a solution spectrum when the solvent is non-polar.

The spectrum of phenol in solution in light petroleum (Figure 11.11) shows two electronic transitions. These correspond to those of benzene in Figure 11.10, but are shifted to 278 and 211 nm. The reason for these shifts to lower energy (longer wavelength) is that there is some delocalization of the π orbitals over the oxygen atom, and this is greater in the excited than in the ground electronic state. The value of $\varepsilon_{max}$ for the $S_1$–$S_0$

**Figure 11.11** Near-ultraviolet absorption spectrum of phenol in solution in light petroleum (after E. A. Johnson, in *UV Atlas of Organic Compounds*, Butterworths, London, 1966, p. D5/4)

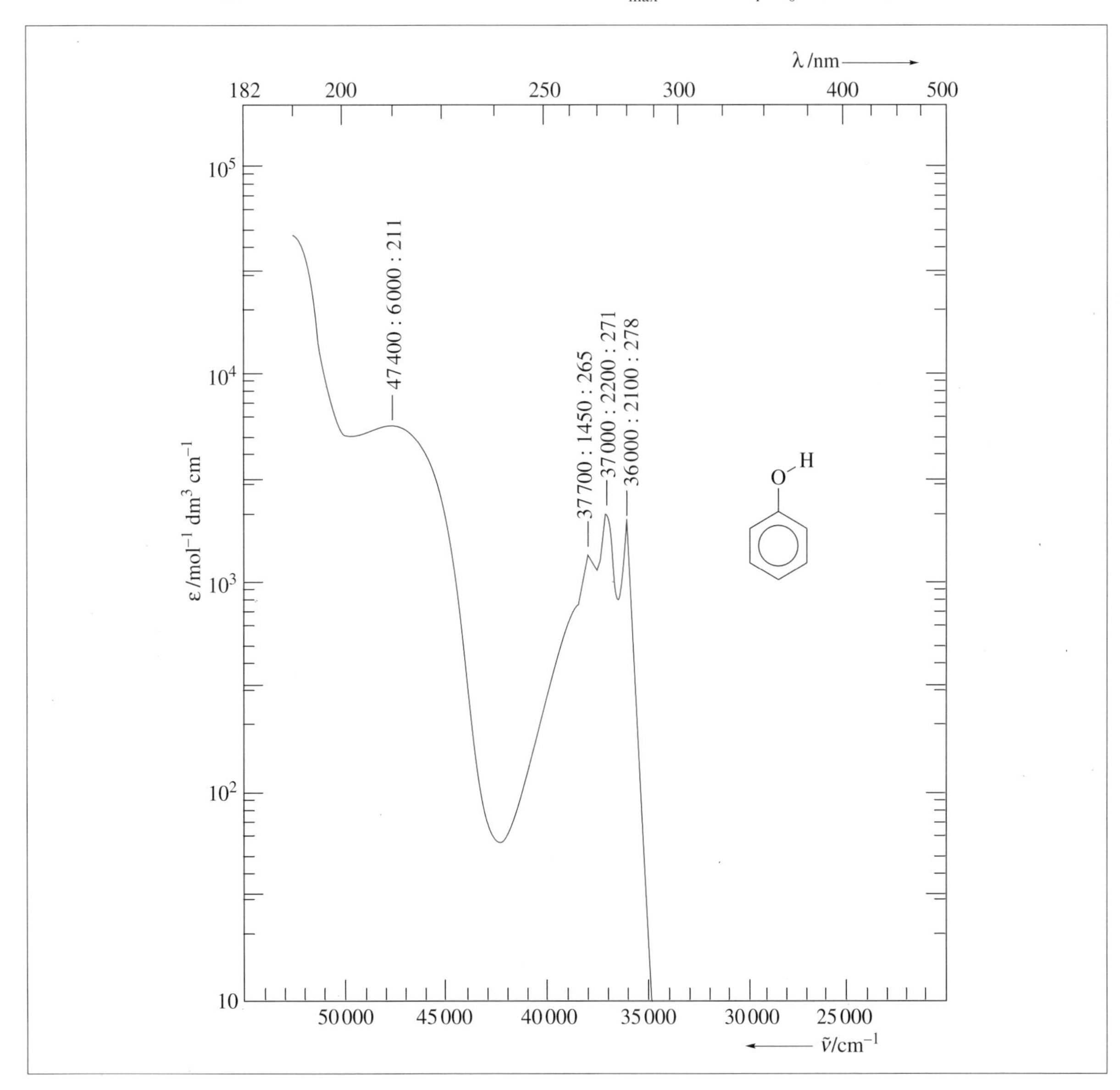

system is $2.2 \times 10^3$ mol$^{-1}$ dm$^3$ cm$^{-1}$, higher than that for benzene, but still low for a $\pi\pi^*$ state. This is due to the fact that, although the electronic transition is allowed, it still retains some of the forbidden character of the corresponding benzene system.

The spectrum of butan-2-one in solution in water (Figure 11.12) is typical of a molecule having a C=O group. The electron promotion involved in the $S_1$–$S_0$ transition, in this case at 267 nm, is from a non-bonding, n, orbital on the oxygen atom to an anti-bonding $\pi^*$ orbital on the C=O group. These orbitals are localized, and are similar to those

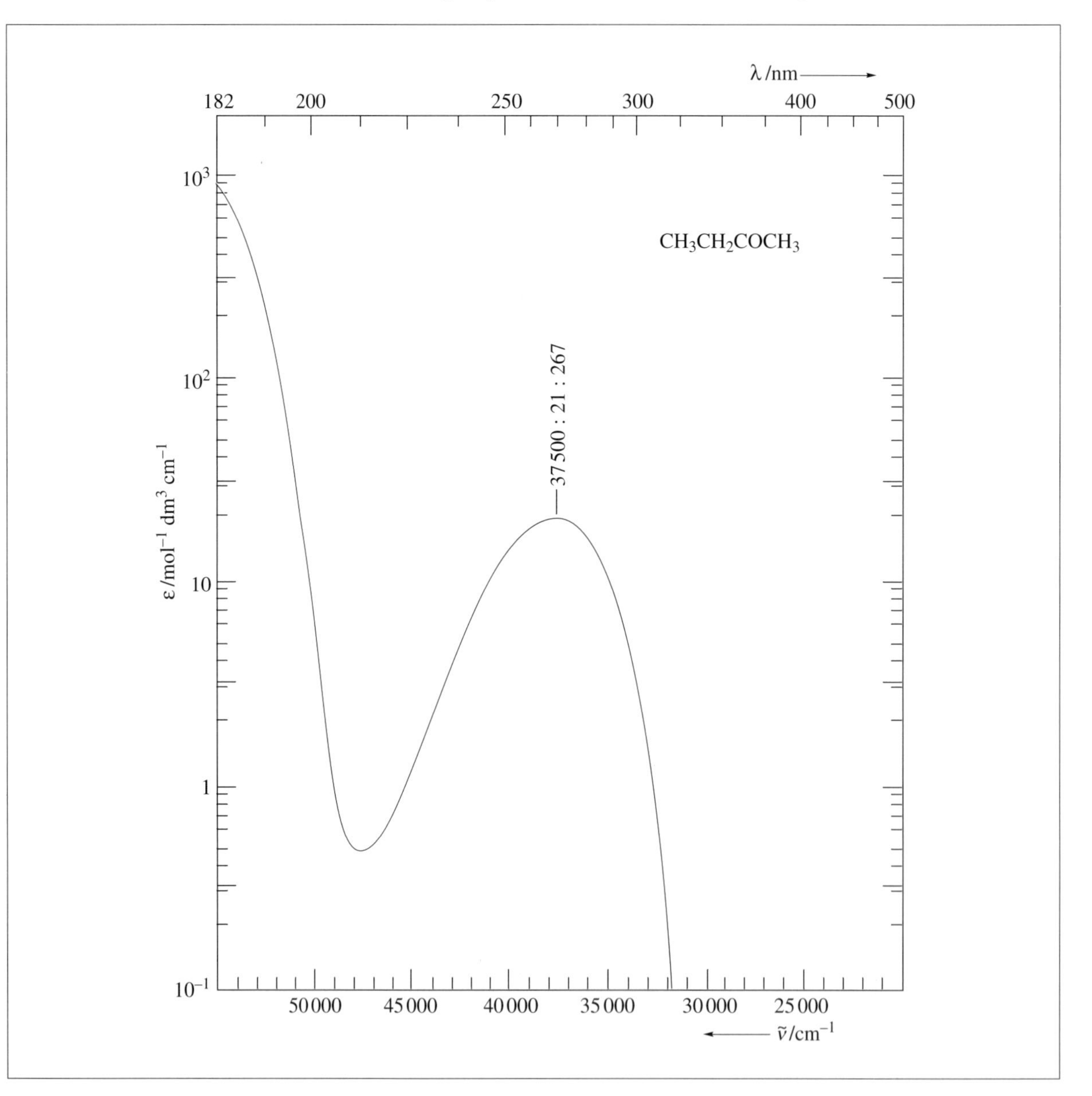


**Figure 11.12** Near-ultraviolet absorption spectrum of butan-2-one in solution in water (after H.-H. Perkampus, in *UV Atlas of Organic Compounds*, Butterworths, London, 1966, p. B1/7)

of methanal (formaldehyde), shown in Figure 5.11; they are little affected by the substitution of the hydrogen atoms by methyl and ethyl groups. The $S_1$ state is an n$\pi^*$ state, as discussed in Section 5.3.2. Such an electronic transition of a molecule containing a C=O group occurs in the region of 275 nm, which is typical for the C=O chromophore. The intensity of the transition is typically very low, having a value of $\varepsilon_{max}$ of only about 20 mol$^{-1}$ dm$^3$ cm$^{-1}$; in the spectrum in Figure 11.12 it is 21 mol$^{-1}$ dm$^3$ cm$^{-1}$.

Although a transition to an n$\pi^*$ state is very weak, it occurs at lower energy than the lowest energy transition to a $\pi\pi^*$ state. (In the spectrum of butan-2-one in Figure 11.12, this latter transition is at a wavelength below 182 nm.) Consequently, it is not likely to be buried under a much more intense transition to a $\pi\pi^*$ state.

We have seen that an electronic spectrum, such as that in Figure 11.12, can be used to indicate the presence of a chromophore, but it may not give much information about any attached groups, in this case the methyl and ethyl groups. The discussion in Section 10.6 shows that vibrational spectroscopy is a more powerful tool for this purpose.

In summary, for the purposes of chemical analysis, vibrational spectroscopy is the more valuable for the identification of a molecule. Electronic spectroscopy is useful for indicating the presence of a chromophore, such as a C=O group or a benzene ring, but its greater strength is in the determination of the concentration of a molecule in solution, when the value of $\varepsilon_{max}$ is known.

## Summary of Key Points

**1.** *Electronic spectra of diatomic molecules*
Selection rules for transitions, between electronic states, in absorption or emission.

**2.** *Vibrational progressions and sequences in spectra of diatomic molecules*
Vibrational selection rule is unrestrictive, but intensity distributions along progressions are governed by the Franck–Condon principle. Sequence intensities depend on the relative populations of the vibrational levels of (in an absorption spectrum) the ground electronic state.

**3.** *Rotational fine structure in spectra of diatomic molecules*
Selection rules for $\Sigma$–$\Sigma$ and $\Pi$–$\Sigma$ transitions. Determination of bond lengths in the two combining electronic states.

**4.** *Vibrational and rotational structure of electronic spectra of polyatomic molecules*
Franck–Condon principle applies to the activity of each vibration. The presence of several progressions and sequences leads to crowded spectra of large molecules. Rotational fine structure of symmetric and asymmetric top molecules are determined by the direction of the electronic or vibronic transition moment.

**5.** *Electronic spectra of molecules in solution*
Solution spectra much easier to obtain than in the gas phase. Very useful in analytical applications. Beer–Lambert law allows the determination of molecular concentration from measurement of the absorbance. Chromophores in a molecule give absorption at a characteristic wavelength and with characteristic intensity.

## Problems

**11.1.** The vibration wavenumbers (the $v'' = 1$–$0$ intervals) for $N_2$ and $Br_2$, in their ground electronic states, are 2356 and 323.6 $cm^{-1}$, respectively. Compare the intensities of the $v' = 1$–$v'' = 1$ sequence bands, in relation to that of the 0–0 bands, in absorption spectra from the ground electronic state, at a temperature of 373.1 K.

**11.2.** In the first excited singlet state, $S_1$, ethyne (acetylene) has a *trans* bent configuration with $\angle$HCC = 120°, compared to 180° in the linear ground state, $S_0$. In addition, the C≡C bond length increases to 1.34 Å (0.134 nm) in $S_1$ from 1.21 Å (0.121 nm) in $S_0$. What can you say about the vibrational coarse structure that you would expect for the $S_1$–$S_0$ absorption spectrum?

**11.3.** What molecular orbitals are involved in the 28 200 $cm^{-1}$ (354.6 nm) band system of *trans*-dimethyldiimide shown in Figure 8.4?

# Further Reading


1. P. W. Atkins and J. de Paula, *Physical Chemistry*, 7th edition, Oxford University Press, Oxford, 2002. (Quantum mechanics, all spectroscopy, symmetry)
2. C. Banwell and E. McCash, *Fundamentals of Molecular Spectroscopy*, 4th edition, McGraw-Hill, London, 1994. (All molecular spectroscopy)
3. J. C. D. Brand, *Lines of Light: the Sources of Dispersive Spectroscopy, 1800–1930*, Gordon and Breach, Amsterdam, 1995. (Historical development of spectroscopy)
4. J. M. Brown, *Molecular Spectroscopy*, Oxford University Press, Oxford, 1998. (Spectroscopy of diatomic molecules)
5. J. R. Ferraro and K. Nakamoto, *Introductory Raman Spectroscopy*, Academic Press, London, 1994. (Raman spectroscopy)
6. J. M. Hollas, *High Resolution Spectroscopy*, 2nd edition, Wiley, Chichester, 1998. (All spectroscopy, symmetry)
7. J. M. Hollas, *Modern Spectroscopy*, 3rd edition, Wiley, Chichester, 1996. (All spectroscopy, symmetry)
8. H. W. Kroto, *Molecular Rotation Spectra*, Dover, New York, 1992. (Rotational spectroscopy)
9. W. Kemp, *Organic Spectroscopy*, 3rd edition, Freeman, San Francisco, 1995. (Spectroscopy for analysis)
10. T. P. Softley, *Atomic Spectra*, Oxford University Press, Oxford, 1994. (Atomic spectroscopy)
11. B. Stewart, *Modern Infrared Spectroscopy*, Wiley, Chichester, 1995. (Infrared spectroscopy for analysis)

# Answers to Problems

## Chapter 1

**1.1.** Using equation (1.1), together with $\lambda$ = 443.5 nm and d$\lambda$ = 443.567 – 443.495 = 0.072 nm, gives the minimum resolving power necessary as $R$ = 443.5/0.072 = 6160.

**1.2.** (a) $l$/cm and $t$/min and (b) $\nu$/kHz and $H$/G.

## Chapter 2

**2.1.** (a) $13.5 \times 10^{-7}$ s = $1.35 \times 10^{-6}$ s = 1.35 μs
(b) $253 \times 10^{-5}$ g = $2.53 \times 10^{-3}$ g = 2.53 mg
(c) $1743 \times 10^{7}$ Hz = $17.43 \times 10^{9}$ Hz = 17.43 GHz
(d) $12.6 \times 10^{-10}$ m = $1.26 \times 10^{-9}$ m = 1.26 nm

**2.2.** (a) (i) $9.748\ 32 \times 10^{9}\ s^{-1}$ = $9.748\ 32 \times 10^{9}\ s^{-1}/2.997\ 92 \times 10^{10}$ cm $s^{-1}$ = 0.325 170 $cm^{-1}$
(ii) $6437.846 \times 10^{-10}$ m = $6437.846 \times 10^{-8}$ cm
$\therefore \tilde{\nu}$ = $1/6437.846 \times 10^{-8}$ cm = 15 533.15 $cm^{-1}$
(b) 12.488 eV = $hc$(8065.54 $cm^{-1}$), as in equation (2.13)
= $6.6261 \times 10^{-34}$ J s × $2.9979 \times 10^{10}$ cm $s^{-1}$ × 8065.54 $cm^{-1}$
= $1.6022 \times 10^{-19}$ J
or $1.6022 \times 10^{-19}$ J × $6.0221 \times 10^{23}$ $mol^{-1}$
= 96.485 kJ $mol^{-1}$

Note that it is important to retain the correct number of significant figures throughout the calculations, corresponding to the number given in the question. This is six figures in (a), seven in (b) and five in (c).

## Chapter 3

**3.1.** From equation (3.13):

$$\begin{aligned} p_\theta &= nh/2\pi \\ &= 5 \times 6.626 \times 10^{-34}\ \text{J s}/2\pi \\ &= 5.273 \times 10^{-34}\ \text{J s} \end{aligned}$$

Note that J s = $m^2$ kg $s^{-2}$ s = $m^2$ kg $s^{-1}$.
Alternatively, since $2\pi$ (= 360°) is the number of radians in a circle, the units of $p_\theta$ may be expressed as J s $rad^{-1}$. In the SI system, the unit of rad may be omitted if clarity is not lost in doing so.
For $n$ = 100, $p_\theta = 1.055 \times 10^{-32}$ J s ($rad^{-1}$)

**3.2.** From equations (3.15) and (3.16):

$$\begin{aligned} m_p v &= h/\lambda \\ \therefore \lambda &= h/m_p v \\ &= 6.626 \times 10^{-34}\ \text{J s}/1.673 \times 10^{-27}\ \text{kg} \times 6034\ \text{m s}^{-1} \\ &= 6.564 \times 10^{-11}\ \text{m} = 65.64\ \text{pm (very small!)} \end{aligned}$$

**3.3.** Equation (3.22) shows that, for f orbitals, $\ell$ = 3. Since $m_\ell$ can take $2\ell + 1$ values, f orbitals are 7-fold degenerate.
For h orbitals (which you will never encounter!), $\ell$ = 5 and they are 11-fold degenerate.

**3.4.** Any table of molar masses of specific isotopes of atoms will tell you that

$$M(^6\text{Li}) = 6.015\ 12\ \text{u}$$

where "u" is the atomic mass unit, and 1 u = $1.660\ 54 \times 10^{-27}$ kg.
Therefore, for one atom:

$$\begin{aligned} m(^6\text{Li}) &= 6.105\ 12 \times 1.660\ 54 \times 10^{-27}\ \text{kg} \\ &= 10.1378 \times 10^{-27}\ \text{kg} \end{aligned}$$

$$\begin{aligned} \therefore \text{mass of nucleus of } {}^6\text{Li}^{3+}, m_{\text{Li}} &= (10.1378 - 3 \times 0.0009) \times 10^{-27}\ \text{kg} \\ &= 10.1351 \times 10^{-27}\ \text{kg} \end{aligned}$$

$$\begin{aligned} \therefore \text{reduced mass of } {}^6\text{Li}^{2+} &= m_e m_{\text{Li}}/(m_e + m_{\text{Li}}) \\ &= [9.109\ 39 \times 10^{-31} \times 10.1351 \times 10^{-27}/(0.0009 \times 10^{-27} + 10.1351 \times 10^{-27})]\ \text{kg} \\ &= 9.108\ 58 \times 10^{-31}\ \text{kg} \end{aligned}$$

From equation (3.4):

Rydberg constant $\tilde{R}$ ($^6Li^{2+}$) $= \tilde{R}_H \times \mu_{Li}/\mu_H$
$= 1.096\ 78\ cm^{-1} \times 10^5 \times 9.108\ 58/9.104\ 43$
$= 1.097\ 28 \times 10^5\ cm^{-1}$

(for the value $\tilde{R}_H$, see Worked Problem 3.2)

From equation (3.10), $\tilde{\nu}(2\text{–}1) = \tilde{R}\,(^6Li^{2+})Z^2(1/1^2 - 1/2^2)$
$= 109\ 728 \times 3^2 \times 3/4$
$= 740\ 664\ cm^{-1}$

Note the inclusion of $Z^2$, where $Z$ is the atomic number, in equation (3.10) for the case when $Z \neq 1$.

## Chapter 4

**4.1.** (a) The ground configuration is

N [He]$2s^22p^3$

Since the only part-filled sub-shell, 2p, is half filled, the innate stability associated with this means that $L = 0$, and the state is an S state. Hund's rule tells us that the ground state has maximum multiplicity, *i.e.* all the spins are parallel and $S = 3/2$. Therefore $2S + 1 = 4$ and the state is a quartet state. When $L = 0$, $J = S$ and therefore the ground state is a $^4S_{3/2}$ state.
(b) The ground configuration is

P [Ne]$3s^23p^3$

As for the nitrogen atom, the only partly filled sub-shell, 3p in this case, is half filled and the three spins are parallel. The ground state is therefore $^4S_{3/2}$, as for nitrogen.
(c) The ground configuration is

Mn [Ar]$3d^54s^2$

Again, the only partly filled sub-shell, in this case 3d, is half filled. Therefore the ground state is an S state. The maximum multiplicity results when all the 3d electron spins are parallel and $S = 5/2$, so that $2S + 1 = 6$. The ground state is therefore $^6S_{5/2}$.

**4.2.** The ground configuration of the nickel atom is

Ni [Ar]$3d^84s^2$

The only sub-shell we have to consider is 3d. This has two vacancies which behave, in respect of $L$ and $S$, like the two electrons in the $3d^2$ ground configuration of titanium. We have seen in Section

4.3.3 that a $d^2$ configuration gives rise to $^1S$, $^3P$, $^1D$, $^3F$ and $^1G$ terms, of which $^3F$ is the ground term. Therefore the ground term for nickel is also $^3F$. Where it differs from titanium is in the fact that the 3d sub-shell in nickel is more than half filled. Consequently, the $^3F_4$, $^3F_3$, $^3F_2$ multiplet is inverted and the ground state is $^3F_4$. This contrasts with the $^3F_2$ ground state of titanium, resulting from a normal multiplet.

## Chapter 5

**5.1.** The bond order is given by one half of the net number of bonding electrons, $b$.
(a) $b = 4$, $\therefore$ bond order = 2.
(b) $b = 2$, $\therefore$ bond order = 1.
(c) $b = 3$, $\therefore$ bond order = 1.5.

**5.2.** The sulfur atom has the AO configuration $[Ne]3s^23p^4$. MOs can be constructed from 3s and 3p AOs in a similar way to those from 2s and 2p AOs, shown in Figure 5.5. The usual rules for MO formation, that the AOs must have the same symmetry and have the same (or similar) energies, are obeyed. The ground MO configuration, for the outer MOs, of $S_2$ is, analogous to that of $O_2$ in equation (5.8):

$$S_2 \quad \ldots (\sigma_g 3p)^2(\pi_u 3p)^4(\pi_g 3p)^2$$

For SO, the rules for forming MOs must still be obeyed. In this case the outer MOs are formed by an LCAO treatment of the 2s and 2p AOs of O and the 3s and 3p AOs of S, the 2s and 3s AOs having similar energies, as do the 2p and 3p AOs. The ground configuration for the outer MOs of SO is:

$$SO \quad \ldots (\sigma 2p,3p)^2(\pi 2p,3p)^4(\pi^* 2p,3p)^2$$

The bond order of both molecules is two. $S_2$ has, like $O_2$, a $^3\Sigma_g^-$ ground state while that of SO is $^3\Sigma^-$; the subscript g does not apply to a heteronuclear molecule.

**5.3.** The vector diagram for Hund's case (c) coupling is shown in Figure A.1. The orbital angular momentum $\boldsymbol{L}$ and the spin angular momentum $\boldsymbol{S}$ are both vector quantities, and are coupled together to give the resultant total angular momentum vector $\boldsymbol{J}$. Comparison with Figure 5.6, which illustrates case (a) coupling, shows that $\boldsymbol{L}$ and $\boldsymbol{S}$ are no longer coupled to the internuclear axis. For this reason, $\Lambda$ and $\Sigma$ are no longer good quantum numbers.

They are replaced by the quantum number $\Omega$ (= 1, 2, 3, ...), which relates to the component $\Omega\hbar$ of the total angular momentum along the internuclear axis. For this coupling case the main label for an electronic state is the value of $\Omega$.

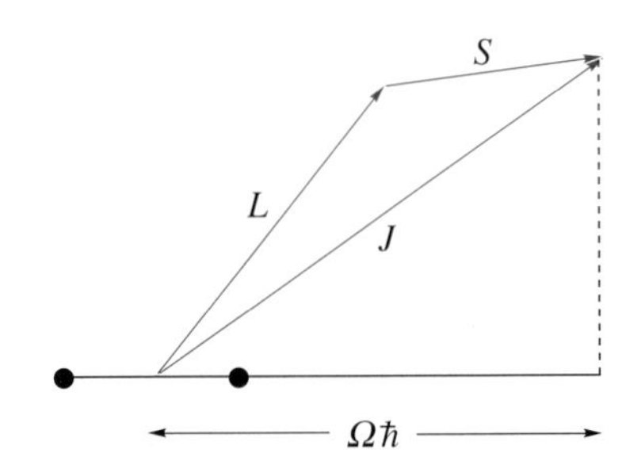


**Figure A.1** Vector diagram for Hund's case (c) coupling

**5.4.** From the ground electron configuration of NO in equation (5.19) we can see that the outer MO configurations of $NO^+$ and $NO^-$ in their ground states are:

$NO^+$ ... $(\pi 2p)^4(\sigma 2p)^2$
$NO^-$ ... $(\pi 2p)^4(\sigma 2p)^2(\pi^* 2p)^2$

All the occupied MOs of $NO^+$ are filled. Therefore the ground state is $^1\Sigma^+$. The electron configuration of $NO^-$ is similar to that of $O_2$, with which it is isoelectronic. Only the g/u classification is lost. The ground state is therefore $^3\Sigma^-$.

**5.5.** The main difference between the MO description of the ground electronic state of hydrogen cyanide and that of ethyne concerns the replacement of a CH group of ethyne with the isoelectronic N atom of hydrogen cyanide. The two electrons on the C–H bond become a lone pair of electrons on the N atom in an orbital which may be described as an sp hybrid.

In pyridine there is a lone pair of non-bonding electrons in an orbital, resembling an $sp^3$ hybrid, attached to the nitrogen atom, and shown in Figure A.2(a). In pyrazine there are similar orbitals attached to each nitrogen atom. Because there are two of them, they may be in-phase or out-of-phase, as shown in Figure A.2(b) or A.2(c), respectively. The additional nodal plane in the out-of-phase MO results in it being at slightly higher energy than the in-phase MO.

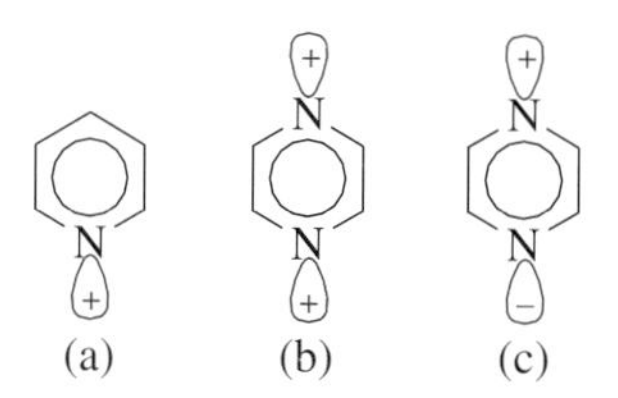


**Figure A.2** Non-bonding, n, MOs of (a) azabenzene (pyridine) and (b, c) 1,4-diazabenzene (pyrazine)

## Chapter 6

**6.1.** The reduced mass, $\mu$, for $^1H^{35}Cl$ is calculated using equations (6.9) and (6.11) to give:

$$\mu = 1.6266 \times 10^{-27} \text{ kg}$$

Then, from equation (6.8):

$$\begin{aligned} v &= (1/2\pi)(515.7 \text{ N m}^{-1}/1.6266 \times 10^{-27} \text{ kg})^{\frac{1}{2}} \\ &= 8.9615 \times 10^{13} \text{ s}^{-1} \end{aligned}$$

$$\begin{aligned} \therefore\ \omega &= 8.9615 \times 10^{13} \text{ s}^{-1}/2.9979 \times 10^{10} \text{ cm s}^{-1} \\ &= 2989 \text{ cm}^{-1} \end{aligned}$$

**6.2.** Putting the data into equation (6.13) gives $G(0) = 1031$ cm$^{-1}$ and $G(1) = 3073$ cm$^{-1}$, with a separation, $G(1) - G(0)$, of 2042 cm$^{-1}$, and $G(10) = 20\,266$ cm$^{-1}$ and $G(11) = 22\,044$ cm$^{-1}$, with a separation, $G(11) - G(10)$, of 1778 cm$^{-1}$.
It is very clear from this result that there is a large reduction of the vibrational interval, due to anharmonicity, as the quantum number $v$ increases.

**6.3.** Putting the data into equation (6.13) to obtain the zero-point vibrational term value, $G(0)$, for each of these molecules gives, for $^1H_2$, $G(0) = 2170.37$ cm$^{-1}$, and, for $^2H_2$, $G(0) = 1542.36$ cm$^{-1}$.

The zero-point level of $^2H_2$ lies considerably deeper, by 628.01 cm$^{-1}$, within the vibrational potential function than that of $^1H_2$. A vibrational potential function, like that in Figure 6.3, is not changed by isotopic substitution; it is the vibrational levels within that potential which are changed. The heavier the isotopically substituted molecule, the closer together are the levels, as exemplified by these zero-point levels. Hence, the dissociation energy, $D_0$, is greater for $^2H_2$ than for $^1H_2$. It follows that the dissociation energy for any X–$^2$H bond is greater than for an X–$^1$H bond. Therefore, if a reaction of a molecule containing such a bond includes the breaking of the bond as a rate-determining step, the reaction rate is reduced for the heavier isotope. In principle, this is true for any isotopic substitution, but the very large difference in the relative masses of $^{2}H$ and $^{1}H$ causes this effect to be particularly pronounced.

## Chapter 7

**7.1.** Since $UF_6$ is a spherical top, all the principal moments of inertia are equal. The easiest way to calculate this is to take any F–U–F line as the inertial axis. Then the moment of inertia, $I$, is given by:

$$I = 4m(\mathrm{F})[r(\mathrm{U–F})]^2$$
$$= 4 \times 18.998 \times 1.6605 \times 10^{-27}\ \mathrm{kg} \times (1.9993 \times 10^{-10})^2\ \mathrm{m}^2$$
$$= 5.0439 \times 10^{-45}\ \mathrm{kg\ m}^2$$

$$\therefore\ B = h/8\pi^2 cI$$
$$= 6.6261 \times 10^{-34}\ \mathrm{J\ s}/8\pi^2 \times 2.9979 \times 10^{10}\ \mathrm{cm\ s}^{-1} \times 5.0439 \times 10^{-45}\ \mathrm{kg\ m}^2$$
$$= 0.055\,499\ \mathrm{cm}^{-1}$$

This value applies to all isotopes of uranium, because the uranium atom is at the centre of mass of the molecule and makes no contribution to the moment of inertia and, hence, to the $B$-value.

**7.2.** For the prolate symmetric top $CF_3I$ we use equation (7.11) and, for the oblate symmetric top $NH_3$, equation (7.12) to calculate the rotational energy levels for $J = 2$ and $K = 0$, 1 and 2. The results are as follows:

| | | | | |
|---|---|---|---|---|
| $CF_3I$ | $K =$ | 0 | 1 | 2 |
| | $F(J,K) =$ | 0.3049 | 0.4451 | 0.8657 |
| $NH_3$ | $K =$ | 0 | 1 | 2 |
| | $F(J,K) =$ | 56.66 | 53.41 | 43.67 |

These results show that the energy levels diverge, with increasing $K$ and constant $J$, for a prolate symmetric top and converge for an oblate symmetric top.

## Chapter 8

**8.1.** From equation (8.1):

$$A = \log_{10}(I_0/I) = \log_{10}(100/15.4) = 0.8125$$

∴ from equation (8.2):

$$c = A/\varepsilon l = 0.8125/10\ 500\ \text{mol}^{-1}\ \text{dm}^3\ \text{cm}^{-1} \times 10.3\ \text{cm}$$
$$= 7.51 \times 10^{-6}\ \text{mol dm}^{-3}$$

**8.2.** To compare the Raman scattering intensities we use equation (8.3) and obtain the ratio of the values of $\lambda^{-4}$ for the two wavelengths, obtained from the two wavenumbers, retaining three significant figures throughout:

| $\tilde{\nu}$/cm$^{-1}$ | $\lambda$/nm | $\lambda^{-4}$/nm$^{-4}$ |
|---|---|---|
| 25 900 | 386 | $4.50 \times 10^{-11}$ |
| 11 900 | 840 | $2.01 \times 10^{-12}$ |

The ratio of these values of $\lambda^{-4}$ is 22.4. This indicates that Rayleigh scattering is about 22 times as intense in the blue compared to the red region, resulting in a cloudless sky appearing blue due to observation of the Rayleigh scattering by the atmosphere.

## Chapter 9

**9.1.** To obtain the best estimate for $B$, measure the distance, in millimetres, between two widely separated rotational lines, for example the $J = 16$ Stokes and anti-Stokes lines; three- or four-figure accuracy is the best that can be achieved. Use the scale, in cm$^{-1}$, to convert the measurement to cm$^{-1}$. This gives a separation of the two $J = 16$ lines of 263.8 cm$^{-1}$. Then, equation (9.8) gives:

$$4B_0(14 + 16) + 12B_0 = 132B_0 = 263.6\ \text{cm}^{-1}$$
$$\therefore\ B_0 = 1.997\ \text{cm}^{-1}$$

This is a surprisingly accurate value, given that the value of $B_e$, the value for equilibrium, is 1.99824 cm$^{-1}$. Centrifugal distortion has been neglected.

**9.2.** From equation (9.3), $B = h/8\pi^2 cI = h/8\pi^2 c\mu r^2$ and, from equation (7.3), $\mu = \frac{1}{2}m(^{14}\text{N})$.

$$\therefore\ r = \{h/8\pi^2 c[\tfrac{1}{2}m(^{14}\text{N})]B\}^{\frac{1}{2}}$$
$$= (6.626 \times 10^{-34}\ \text{J s}/8\pi^2 \times 2.998 \times 10^{10}\ \text{cm s}^{-1} \times 7.002\ \text{u} \times 1.661 \times 10^{-27}\ \text{kg} \times 1.997\ \text{cm}^{-1})^{\frac{1}{2}}$$
$$= (1.205 \times 10^{-20}\ \text{m}^2)^{\frac{1}{2}}$$
$$= 1.098 \times 10^{-10}\ \text{m (or 1.098 Å)}$$

**9.3.** Measuring the separation of the $J = 6$–5 and 12–11 lines, using the scale on the figure, gives $12B = 45.62$ cm$^{-1}$ and $B = 3.802$ cm$^{-1}$.

## Chapter 10

**10.1.** As stated in Section 10.3, the O and S branches have $\Delta J$ = –2 and +2, respectively. The O(10) and S(10) transitions are separated by $(18 \times 4B) + (2 \times 6B) = 84B$, giving $B = 1.907$ cm$^{-1}$. Using the same method as in Worked Problem 10.1, the reduced mass $\mu$ = $1.139 \times 10^{-26}$ kg, and $r = 1.13(5) \times 10^{-10}$ m = 1.13(5) Å.

The approximations involved are that $B$ has been assumed to be the same for both vibrational states, $v$ = 0 and 1, and that centrifugal distortion has been neglected.

**10.2.** The $CO_2$ molecule is linear (O=C=O) and has three vibrations: the symmetric C–O stretching, the asymmetric C–O stretching and the bending vibration. The asymmetric stretching and bending vibrations involve a change of dipole moment and therefore are infrared active. The symmetric stretching vibration does not involve a change of dipole moment. There is, however, a change of polarizability, so the vibration is Raman active. Because the molecule has a centre of symmetry, the infrared and Raman active vibrations are mutually exclusive.

**10.3.** Two types of C–H stretching vibration are expected, those within the $CH_3$ and the CH groups. According to the information in Table 10.1, the former, of which there are three, should be in the region of 2960 cm$^{-1}$ (observed at 2941 cm$^{-1}$) and the latter in the region of 3300 cm$^{-1}$ (observed at 3335 cm$^{-1}$). The C–C and C≡C stretching vibrations are expected in the region of 900 cm$^{-1}$ (observed at 931 cm$^{-1}$) and 2050 cm$^{-1}$ (observed at 2142 cm$^{-1}$), respectively.

## Chapter 11

**11.1.** Using equation (9.11) to obtain the population ratio of the $v'' = 1$ to that of the $v'' = 0$ level:

For $N_2$, $$hc/kT = 6.626 \times 10^{-34}\ \text{J s} \times 2.998 \times 10^{10}\ \text{cm s}^{-1}/1.381 \times 10^{-23}\ \text{J K}^{-1} \times 373.1\ \text{K}$$
$$= 3.855 \times 10^{-3}\ \text{cm}$$

$$\therefore N_1/N_0 = \exp(-3.855 \times 10^{-3}\ \text{cm} \times 2356\ \text{cm}^{-1})$$
$$= 1.137 \times 10^{-4}$$

Similarly, for $Br_2$:

$$N_1/N_0 = \exp(-3.855 \times 10^{-3}\ \text{cm} \times 323.6\ \text{cm}^{-1})$$
$$= 0.2872$$

Consequently, the intensity of the sequence band is much higher, by a factor of about 2500, in $Br_2$ than in $N_2$.

**11.2.** The two major geometry changes of ethyne from $S_0$ to $S_1$, the decrease of the ∠HCC and the increase in the C≡C bond length, result in fairly long progressions in the vibrations $\nu_4$ (the *trans* bending vibration) and $\nu_2$ (the C≡C stretching vibration), respectively. These vibrations are illustrated in Figure 6.6.

**11.3.** Figure 8.4 shows that the 354.6 nm electronic band system of *trans*-dimethyldiimide is very weak, with $\varepsilon_{max} \approx 14$ mol$^{-1}$ dm$^3$ cm$^{-1}$. This is characteristic of a transition to an nπ* $S_1$ state in which an electron is promoted from a non-bonding, n, orbital on the nitrogen atoms to a π* orbital in the N=N group. The n orbital resembles that in either Figure A.2(b) or A.2(c), an in-phase or out-of-phase combination of $sp^2$ hybrid orbitals on the nitrogen atoms. Consideration of which of these is involved in the 354.6 nm system is beyond the scope of the present discussion. The π* orbital resembles that for ethene shown in Figure 5.10.

# Subject Index